WISCONSIN LOCAL FOODS JOURNAL

2014

Praise for past Wisconsin Local Foods Journals

"It's easy to eat local in the summer. But what about the rest of the year? Terese Allen and Joan Peterson have the answers for you. And they're packaged handsomely in a handy new spiral-bound book."
—Nancy Stohs, Milwaukee Journal-Sentinel

"A feast for the eyes and the go-to source for getting organized about enjoying Wisconsin's bounty. Love it!"
—Jamie Lamonde, Edible Madison Magazine

"Through this journal the abundance of local eating in Wisconsin finds 365 days to celebrate delicious."
—Odessa Piper, Founder, L'Etoile Restaurant

"There aren't more eloquent Dairyland patriots than Joan Peterson and Terese Allen, who together they do full poetic justice to [Wisconsin's] bulging harvest."
—Raphael Kadushin, The University of Wisconsin Press

"Show you how to eat deliciously—and sustainably—all year long."
—Judith Fertig, *Heartland: The Cookbook*

"Looks good, tastes good, does good. By two of Madison's most food-forward thinkers, [the Journal has a] logical and easy-to-use format."
—Mary Bergin, *Hungry for Wisconsin* and *Eat Smart in Germany*

"A great resource for cooks and eaters….Teaches us how to start with our finest ingredients and learn to cook with what we have."
—Matt Feifarek, Slow Food Madison

"Locavores need connections to food growers, and that process just got easier…[with] the *Wisconsin Local Foods Journal*, a helpful resource for Wisconsin eaters."
—Amy Lou Jenkins, Milwaukee Green Living Examiner

"My eyes savor the beautiful illustrations, my mouth the flavorful seasonal recipes, and my brain says it's the right thing to do, for myself, my family, the farmers, and my community."
—Bruce Dethlefsen, Wisconsin Poet Laureate (2011-2012)

"A unique culinary guide that provides recipes and practical tips on how to make the most of seasonal garden bounty on a daily basis."
—Susan Troller, Wisconsin State Journal

WISCONSIN LOCAL FOODS JOURNAL

SUSTAINABLE EATING ALL THROUGH THE YEAR

2014

CHEESE AND CHEESEMAKERS EDITION

by Joan Peterson & Terese Allen

To my son-in-law Edward, the family cheesemaker, who introduced me
to many of the artisanal cheeses we celebrate in this edition of the
Wisconsin Local Foods Journal.
—Joan

To Susie/Rose/Sueallen, who can hit the broad side of a barn
but can't eat cheese. I like your pink glasses.
—Terese

Acknowledgments

We are most grateful for the assistance of the following people: Susan Chwae, creative designer and technical expert, who made everything fall nicely into place. Wendy Allen, copy editor who gave freely of her time to search for pesky errors. Jeanne Carpenter, cheesehead extraordinaire, who wrote a compelling essay about how Wisconsin reinvigorated its status as America's Dairyland, provided outreach assistance, and, along with her husband Uriah Carpenter, generously provided photos. Mary Bergin and Bill Lubing, fellow journalists, for bighearted project support and use of their photos. Doris Baier, who frequently plied us with helpful articles and tidbits about the local foods scene around the state. REAP Food Group staff members Miriam Grunes, Conor Moran, Jessica Wetzel, and Theresa Feiner for production and outreach assistance, and all the dedicated REAP volunteers who help with this project. As of this writing we don't know who all of them will be, but we'd like to tip our hats now to several who were particularly helpful last year: book reps Liz Chapa, Pamilyn Hatfield, and Debra Shapiro, and "Mr. Delivery," Willie Goehrig.

We'd also like to make special note of the Wisconsin Milk Marketing Board (WMMB) and Organic Valley/CROPP Cooperative for generous funding support, content contributions, and use of photos, and thank the good-guy liaisons from both organizations who encouraged and aided us: Heather Porter Engwall and Greg Long of the WMMB, and Jamie Lamonde and Vicky Reeves of Organic Valley.

We're ever-grateful to our husbands, David Peterson and Jim Block, for keeping our "life support" running at full tilt. And finally, we'd like to thank and salute all the talented and resourceful Wisconsin cheesemakers who shared with us their stories, photos—even cheese jokes—and who craft the yummy, award-winning traditional and artisanal cheeses that continue to bring fame to our state.

Joan Peterson and Terese Allen

Table of Contents

Recipes

Introduction

Welcome to the *Wisconsin Local Foods Journal*, an "every day" guide to sustainable eating. The featured topic for this, the third annual edition of the book, is cheese. If there's a more beloved local food in this state, we don't know what it is!

This is a multi-purpose book, to be used in any or all of the following ways:

- As an engagement or desk calendar;
- As a shopping guide to what's in season throughout the year, and the cheeses to pair with it (see What's in Season pages);
- As a cookbook with dozens of recipes that feature the best of Dairyland cheese;
- As a travel resource to help you locate cheese factories and cheese shops across Wisconsin (see the maps and listings on pages 173–190);
- As a guide to buying, storing, and pairing cheese (pages 171–172), and to cheese varieties and terms (pages 161–170);
- As a source of inspiration and guidance from top Dairyland cheesemakers (see "Featured Cheesemaker" sections);
- As a place to find Dairyland discounts and special offers (see Cheese Coupons section, pages 191–194);
- As a means to extend your commitment to creating a more regionally focused, sustainable food system (see About REAP Food Group, page 199, and 2014 REAP Events, page 200.

We are very excited that this year's Journal is sponsored by two of the nation's premier dairy organizations: the Wisconsin Milk Marketing Board and Organic Valley (see pages 195–196). Their generous backing allowed us to produce what we believe is the most beautiful and useful journal to date. More importantly, it supports crucial sustainable food efforts, because all proceeds from the book go to REAP Food Group, whose initiatives include farm to school programs, the annual *Southern Wisconsin Farm Fresh Atlas*, and the farm to business project called Buy Fresh Buy Local.

We're also tickled this year to include an essay by artisan cheese authority Jeanne Carpenter, founder of Wisconsin Cheese Originals and the gifted blogger at CheeseUnderground.com (as well as our nominee for cheese laureate of the state, should there ever be one). Jeanne's good news about the "state of dairy" can be found on pages 11–13.

We hope this book will be a helpful, day-to-day tool for all cheeseheads who want to eat more deliciously, healthfully, and sustainably.

From our cheese boards to yours,
Joan Peterson and *Terese Allen*

Can you say cheese? ©Wisconsin Milk Marketing Board, Inc. (Digitally altered)

Taking Back America's Dairyland

By Jeanne Carpenter

Say "Wisconsin" and instant images come to mind: construction-orange blocks of Cheddar, black- and white-spotted bovines, and green and gold Packer fans in foam cheese hats. It's a fact: no other state is so associated with cheese and cows. With more than 100 years of dairying and cheesemaking, it seems Wisconsin has always been known as "America's Dairyland."

That's why it's hard to believe that only a decade ago, the Wisconsin dairy industry was in retreat. Long regarded as the backbone of the state's economy, Wisconsin dairy was reeling from falling farm numbers, decaying processing plants, declining profits, and little enthusiasm. While mega dairies sprouted up in western states with cheaper land, the price of Midwest milk plummeted, and the farming lifestyle that defined the entire state was in such peril that many farmers and cheesemakers—despite generations of tradition—considered leaving the business.

In short, an infrastructure once built to support a thriving industry was crumbling. Innovation was in short supply. So how, in less than a generation, did Wisconsin's dairy community transform a seemingly doomed enterprise into not only a success story, but also a model for sustainable family farms and specialty cheese plants across the nation?

It's simple: the state of Wisconsin stopped taking dairy for granted.

In 2004, a consortium of partners and industry leaders launched an ambitious effort to reinvigorate the Wisconsin dairy industry. As a recently-hired spokesperson at the Wisconsin Department of Agriculture, I was assigned to write up goals of what was a $2 million initiative to reclaim Wisconsin's title of "America's Dairyland."

I had a vested interest in this project, being a former farm kid who, growing up, had watched not only my father but all of our neighbors sell the family dairy herd and convert our dairy barns into steer sheds. The project was to become known as the "Value Added Dairy Initiative." Its architect, Will Hughes, put together an aggressive set of five-year goals that was published in 2004:

1. Increase milk production by 15 percent
2. Increase specialty cheese volume by 25 percent
3. Create 50 new value-added dairy enterprises
4. Increase total economic output by $1.5 billion

At the time, I was too naïve to realize that most of these goals were unattainable in five years. You have to understand that at the time, Wisconsin was losing three dairy farms a day. The papers were full of auction notices. By the end of 2003, Wisconsin was home to 15,904 dairy farms, down from nearly 50,000 in 1980. Milk production reflected the loss of farms, falling to 22.3 billion pounds in 2003. The decline of cheese plants followed, retreating to fewer than 115 for the first time in more than a century. Farmers, processors, and cheesemakers across the nation watched Wisconsin's dairy demise as media reports predicted a California takeover of U.S. cheese production. At the Department of Agriculture, I was assigned to prepare a news release conceding Wisconsin's title of America's Dairyland to those damn happy cows in California.

But a funny thing happened. With lofty goals, the promise of support from key state organizations, and a team of optimistic young bureaucrats traveling the state with the motto of "We're from the government and we're here to help you," a mindset that had been dark and gloomy began to change to one that was enthusiastic, innovative, and focused on growth. We knew that if successful, the effort had the potential to strengthen and help prosper all sectors of Wisconsin dairy, including individual farm families, dairy processors, state, national, and international markets, and finally, rural communities.

Early strategies were developed to address the specific goals of the program, including developing the Dairy Business Innovation Center to deliver needed technical services to dairy farms and cheese plants, and implementing programs to stimulate investment, including tax credits and job training programs. Partners joining the Wisconsin Department of Agriculture's effort included such major dairy organizations as the Wisconsin Milk Marketing Board and the Wisconsin Cheese Makers Association; and educational systems, including the University of Wisconsin and the Wisconsin Center for Dairy Research.

Often, big dreams have big results. This was one of those times. Within five years, the initiative's successes were astounding. We passed the once-considered "lofty" five-year goals and rewrote them over and over, only to watch our state's dedicated dairy farmers and cheesemakers smash them again and again. As early as 2005, Wisconsin's milk production began to climb and today totals three billion more pounds than in 2003. The percentage of high quality, high value specialty cheese today totals more than 586 million pounds, almost 48 percent of U.S. production, and up from 302 million pounds in 2003.

Cheese varieties produced in Wisconsin now total more than 600, up from 300 just ten years ago, and countless new products have been introduced or are under development. In addition, the 20-year trend of declining dairy plants was reversed, with 129 cheese plants now calling Wisconsin home. Finally, milk prices paid to Wisconsin farm families steadily increased. During six of the eight years of the core "Value Added Dairy Initiative" years, milk prices averaged above $15 per hundredweight, compared with only one year in the previous eight.

As you can imagine, all of this industry growth led to an increased economic impact. In 2004, Wisconsin dairy contributed $18.5 billion to the state. Today, it is responsible for $26.5 billion in economic activity.

While the numbers are impressive, they still do not tell the entire story. Many unanticipated results occurred, and some remain under development. For example, the state's recommitment to dairy helped make changes and upgrades in regulatory and licensing programs to ensure that optimism and innovation in the dairy industry would continue. In some cases, regulatory requirements that dated to the 1920s were upgraded. In addition, a new generation of dairy farmers, cheesemakers, and leaders are coming forward, as the industry offers growth and excitement.

In short, every original goal set by a team of wide-eyed dreamers in 2004 was met, and in most cases, exceeded. I am honored to have had the unprecedented privilege of sitting in hundreds of dairy farm kitchens and milking parlors and standing around cheese vats, encouraging farmers and cheesemakers—young and old—to reclaim their place in history. It is because of their dedication that, once again, Wisconsin is indeed "America's Dairyland."

Photo by Uriah Carpenter.

Jeanne Carpenter is a cheese geek, but she didn't start out that way. Raised on Velveeta on a family farm in Wisconsin, after college she worked as a journalist, a corporate project manager, and a spokesperson for the Wisconsin Department of Agriculture. In 2007 she started a public relations company to promote artisan cheese, and in 2009 launched Wisconsin Cheese Originals, a member-based organization dedicated to tasting and learning about Wisconsin artisan cheeses. She leads a variety of events, including cheese classes, tours, cheesemaker dinners, and the annual Wisconsin Cheese Originals Festival. Since 2006, she has blogged on CheeseUnderground.com, giving cheese-starved readers everywhere the inside scoop on America's Dairyland.

Winter is for fondue. Photo by Terese Allen.

Try not to drool as you read the following! We asked craft cheesemakers to name a specialty of theirs that would be ideal for winter dining and to match it with food or drink.

BRUCE WORKMAN, EDELWEISS:
"***Emmentaler***. I love to make fondue and this is the perfect cheese for it. Cold weather, bottle of wine and a pot of fondue: Life doesn't get much better."

KEN HEIMAN, NASONVILLE DAIRY:
"***Pepper Jack***—this cheese really adds some much-needed heat in the cold months. It's great on its own, in a grilled cheese, or mixed in a bowl of chili."

MARIEKE PENTERMAN, HOLLAND'S FAMILY CHEESE:
"[***Gouda*** with] clove with some good ale beer and good salami."

KATIE HEDRICH, LACLARE FARMS:
"***Evalon***. I love this cheese in the winter because I love using it anywhere Parmesan is used. It also goes really well with a dry red wine, which I really enjoy in winter."

VEGETABLES (INCLUDES HYDROPONIC & STORED)

- [ ] Arugula
- [ ] Beets
- [ ] Cabbage
- [ ] Carrots
- [ ] Celeriac
- [ ] Dried beans
- [ ] Dried chiles
- [ ] Garlic
- [ ] Lettuce
- [ ] Microgreens
- [ ] Mushrooms (cultivated)
- [ ] Onions
- [ ] Parsnips
- [ ] Potatoes
- [ ] Rutabaga
- [ ] Shallots
- [ ] Spinach
- [ ] Sprouts
- [ ] Sweet potatoes
- [ ] Tomatoes
- [ ] Turnips
- [ ] Winter radishes
- [ ] Winter squash

FRUITS

- [ ] Apples
- [ ] Frozen fruits
- [ ] Pears

MEATS & FISH

- [ ] Beef
- [ ] Bison
- [ ] Chicken
- [ ] Emu
- [ ] Lamb
- [ ] Panfish
- [ ] Pork
- [ ] Rainbow trout
- [ ] Sausages
- [ ] Smoked fish
- [ ] Turkey
- [ ] Venison
- [ ] Whitefish

EGGS & DAIRY

- [ ] Butter
- [ ] Buttermilk
- [ ] Cheese (cow, goat & sheep milk)
- [ ] Cottage cheese
- [ ] Cream
- [ ] Curds

- [ ] Dips
- [ ] Eggs
- [ ] Ice cream
- [ ] Milk
- [ ] Yogurt

OTHER

- [ ] Baked goods
- [ ] Beer
- [ ] Breads
- [ ] Cider
- [ ] Flours
- [ ] Herbs
- [ ] Hickory nuts
- [ ] Honey
- [ ] Horseradish
- [ ] Jams
- [ ] Maple syrup
- [ ] Pesto
- [ ] Pickles & preserves
- [ ] Popcorn
- [ ] Prepared foods
- [ ] Salsa
- [ ] Sauces
- [ ] Spirits
- [ ] Sunflower oil
- [ ] Tortillas
- [ ] Vinegars
- [ ] Wild rice

Cheese of the Month: Brick

Profile: Brick cheese was invented in Wisconsin in 1877 by Swiss-born John Jossi, a Limburger factory operator. He created a firmer, more sliceable, and somewhat less powerful cheese than Limburger. Both types are traditionally surface-ripened, washed-rind cheeses, but Jossi shaped the new variety into blocks and weighted it with Bricks—thus,  the name. To this day some people call Brick the "married man's Limburger," although a well-aged Brick is no slouch in the stinky cheese department. Most Brick today is not surface-ripened and is sold young and quite mild; it's a great melter and can be substituted for mozzarella or similar mild, melting cheeses in recipes. Funky, potent, aged Brick, on the other hand, is for adventuresome eaters only. (Both are made from pasteurized cow milk.) If you want to impress friends in other parts of the country, where classic Brick is rare, give them old-style Brick; it's a real taste of Wisconsin adventure.

Wisconsin-made styles include: Washed-rind (traditional) and mild Brick (commercial).

Taste/Aroma: Young Brick is mild, sweet, and a little nutty, but as the cheese ages it gains pungency, tang, a little spice, a little gaminess, and a lot of savor. It also gains aroma—surface-ripened, aged Brick has a strikingly funky odor.

Texture: Semi-firm, supple, sliceable. Can have a pleasant hint of nubbiness due to the pressed curds that form the cheese.

Go-withs: Pickles, mustard, raw onions, whole-grain bread, liverwurst, asparagus, ham, caraway, cabbage, tomatoes, pickled herring, apples.

What to drink: Pinot Noir, Beaujolais, fruity white wines, or red ales with mild Brick; Riesling, Chardonnay, pale ale, bock, porter, or stout with aged Brick; apple cider with either.

Classic dishes: Liverwurst and Brick sandwich with mustard and pickles; grilled cheese sandwich.

Try this: Slice garlicky summer sausage and skillet-fry it in a little butter, then layer with aged Brick and thinly sliced red onions inside a pumpernickel roll. Got horseradish mustard? Slather away.

January 2014

Wednesday 1

New Year's Day
New Moon

Thursday 2

Friday 3

Saturday 4

Cheddar Sage Biscuits

Servings: 12

Winter is a welcome thing…when your kitchen is filled with the warm aroma of cheese biscuits, that is. For this recipe, grating frozen butter is a convenient way to incorporate small, equal-sized bits of very cold fat into flour, just what is called for to help give biscuits a high rise. Likewise, keep the cheese chilled once you've grated it; if it gets too soft, it can clump up in the dough.

> 2 cups unbleached high-protein flour (such as King Arthur)
> 1 tablespoon chopped fresh sage or 1 ½ teaspoons dried
> 2 teaspoons baking powder
> 1 teaspoon baking soda
> ¾ teaspoon salt
> 3 tablespoons frozen unsalted butter (in one chunk)
> 1 ½ cups chilled coarsely grated aged Cheddar
>   (about 2 ½ ounces), divided
> ½ cup plain yogurt
> ½ cup whole or two-percent milk

Heat oven to 400 degrees. Line a baking sheet with parchment paper. (If parchment paper isn't available, use an ungreased baking sheet.)

Whisk flour, sage, baking powder, baking soda, and salt in a bowl. Take the butter directly from the freezer and leave the wrapper on one end. Using the largest holes on a hand grater, grate the frozen butter directly into the flour mixture. Whisk lightly to distribute the butter evenly throughout the flour. Add 1 cup of the chilled grated cheddar and whisk again.

Using another bowl, stir yogurt and milk together until smooth. Add this to the dry ingredients and stir to combine, taking care not to over mix. You want the ingredients to just come together into a sticky dough at this point. Flour your hands and gently roll the dough between them into 12 rough-surfaced balls, again without overworking the dough. Place the dough balls 1 to 2 inches apart on the baking sheet. Sprinkle each one with a little of the remaining cheese. Bake until brown-tipped and a little crusty, 15–20 minutes. Serve hot or warm.

January 2014

Day	Date
Sunday	5
Monday	6
Tuesday	7
Wednesday	8
Thursday	9
Friday	10
Saturday	11

Sweet Potato and Swiss Cheese Soufflés

Servings: 6–8 small individual soufflés

*This is the season when the oven is on a lot. The next time yours is
heating up, throw in a sweet potato so that you'll have it cooked to make
this dish. Adapted from an online recipe, it very successfully marries
the sweetness of the root vegetable with the nuttiness of aged Swiss. We
recommend something like the Nutty Swiss from Bleu Mont Dairy or the
award-winning Grassfed Emmentaler from Edelweiss Creamery.*

    1 tablespoon butter, plus a little for the baking cups
    3 tablespoons minced shallots
    1 teaspoon minced garlic, mashed to a paste with flat of knife
    2 teaspoons chopped fresh thyme or 1 teaspoon dried
    ¼ teaspoon cayenne
    salt and pepper
    1 tablespoon flour
    ½ cup whole or two-percent milk
    1 heaping cup coarsely grated aged Swiss cheese, divided
    1 cup cooked sweet potato purée
    3 large eggs, separated and then brought to room temperature

Heat oven to 375 degrees. Butter 6–8 small soufflé cups or ramekins.
Place on baking sheet. Melt 1 tablespoon butter in a saucepan over a
medium-low flame. Add shallots, garlic, thyme, and cayenne, plus a
little salt and pepper. Cook, stirring frequently, until shallots are tender.
Stir in flour and keep cooking, stirring often, a few more minutes.

Whisk in the milk a little at a time and keep stirring until mixture
is thickened, 2–3 minutes. Remove from heat and whisk in ⅓ cup
cheese. Whisk in another ⅓ cup cheese, then the sweet potato purée,
and finally the egg yolks. Transfer to a large bowl and cool to room
temperature.

Using electric beaters or a clean whisk, beat egg whites and a pinch of
salt until stiff peaks form. Whisk ¼ of the beaten whites into the sweet
potatoes. Switch to a rubber spatula and gently fold in remaining egg
whites. Divide mixture into buttered baking cups and sprinkle with
remaining cheese. Bake until puffed and barely set, 17–20 minutes.
Serve *pronto*.

January 2014

Sunday	12
Monday	13
Tuesday	14
Wednesday	15
Thursday	16
Friday	17
Saturday	18

Spiedini Linguine
Servings: 3–4

Here's a simple dish that utilizes ingredients many cooks keep on hand in the kitchen, making it perfect for those bone-snapping winter evenings when the last thing you want to do is make a trek to the grocery store. Locally sourced items in the mix include hard cheese, eggs, pesto, and fresh pasta. (Who says you can't find local ingredients in Wisconsin during winter?)

The recipe was inspired by breading that was leftover from preparing spiedini (Sicilian kebobs) one night—thus the name. Mixed with eggs and cheese, the breading is browned in olive oil until it looks—and satisfies— like cooked Italian sausage. Almost any type of Italian-style premium hard cheese will work with it, such as Asiago, Romano, or Parmesan. Top pick? That's a tough one, but perhaps it would be the nutty-sweet BelleVitano Gold from Sartori Cheese.

    1 cup semolina or panko-style bread crumbs
    ½ – ⅓ cup coarsely grated Italian-style hard cheese
    ¼ cup basil pesto
    2 eggs, beaten
    3 tablespoons extra virgin olive oil, plus additional oil for garnish
    1 package (9 ounces) RP's Pasta whole wheat linguine or other
       fresh linguine
    salt and pepper

Bring a large pot of salted water to boil. Warm an oven-proof bowl in a 200-degree oven.

Meanwhile, mix bread crumbs, cheese, pesto, and eggs in a bowl. Heat a heavy skillet over a medium high flame for several minutes. Add 3 tablespoons olive oil and swirl to coat the bottom of the skillet. Stir in the bread crumb mixture and sauté it, stirring and "crumbling" it as you would Italian sausage, until well-brown and crispy, about 10 minutes.

Stir the linguine into the boiling water until the pasta strands are separated. Boil until barely tender, 2–3 minutes. Drain pasta. Toss with crumb mixture plus salt and pepper to taste. Serve immediately, adding a drizzle of additional olive oil to each serving.

January 2014

| Sunday | 19 |
| Monday | 20 |

Martin Luther King Jr. Day

Tuesday	21
Wednesday	22
Thursday	23
Friday	24
Saturday	25

Gary Grossen, Master Cheesemaker

Babcock Hall Dairy Plant
1605 Linden Drive, Madison
608-263-3215
babcockhalldairystore.wisc.edu

Gary Grossen grew up in a cheese factory…up*stairs*, that is. His folks owned Prairie Hill Cheese Factory in Monroe, and the family apartment was above the factory. At an early age he learned that cheesemaking was hard but very rewarding work. Gary eventually joined his father in running the plant and in 1990 took over the business. The business was sold in 2001, but Gary still managed it until he became the cheesemaker in 2005 at the Babcock Hall Dairy Plant at UW-Madison. Now he mentors students aspiring to be cheesemakers themselves and supervises plant production. Gary holds Wisconsin Master Cheesemaker certificates in Cheddar, Muenster, Brick, Gouda, and Havarti cheeses.

What is your company's history? [It started with] a milestone in modern dairying—the development of a simple and accurate measure of the butterfat content of milk. University of Wisconsin biochemist Stephen M. Babcock developed the test in 1890. It made him internationally famous and revolutionized milk production and marketing. The test provided a rational basis of milk evaluation, and prompted better breeding, feeding, and milk production practices. Babcock instructed dairy farmers in the use of the test, which led to the start of the nation's first dairy manufacturing short course. In 1951 Babcock Hall Dairy was established in its current location. Cheese has been produced at the plant for many years. The real purpose at Babcock Hall Dairy is the training and education of students and others who are working in the fields of the dairy industry.

Please describe your cheeses. At Babcock Dairy we manufacture Havarti, Pesto Havarti, Dill Havarti, Marble Jack, Marble Jack with Chives, Monterey Jack, Monterey Jack with Chives, Romano, Cheddar, Gouda, Juustoliepa, Juustoliepa with Peppers, Swiss, Brick, and Colby.

Which of the cheeses you make is your favorite? Gouda and Swiss. I love making cheeses that produce eyes (holes). Both are challenging, but Swiss is the most rewarding because there are so many different factors that can affect eye formation.

Where do you get your inspiration to craft a new cheese? From all the new and different starter cultures and adjuncts that are available for developing a specific flavor in order to craft a new cheese. It is very interesting, and for a cheesemaker it's the challenging aspect.

January 2014

Sunday	**26**
Monday	**27**
Tuesday	**28**
Wednesday	**29**
Thursday	**30**
	New Moon
Friday	**31**
	Chinese New Year
Saturday	**1**

Mountain of goodness. ©Wisconsin Milk Marketing Board, Inc.

When asked, "What do you think will be the next rage in cheese or in cheese and beverage pairing?", here's what three cheese connoisseurs predicted:

FRANCIS WALL, BELGIOIOSO:

"***Burrata***—this fresh cheese resembles a fresh mozzarella ball, but when split open, [there is] a rich-tasting, soft filling of mozzarella and heavy cream inside."

PAULA HOMAN, RED BARN FAMILY FARMS:

"I think beer will perhaps get more attention as a pairing for cheese, but I also like the idea of pairing cheeses with interesting artisan breads."

SUE MERCKX, SARTORI CHEESE:

"Artisan beers and cheese, and cordials and cheese, seem to be on trend. There was even an article recently that called out soda pairings with cheese."

VEGETABLES (INCLUDES HYDROPONIC & STORED)

- ☐ Arugula
- ☐ Beets
- ☐ Burdock root
- ☐ Carrots
- ☐ Celeriac
- ☐ Dried beans
- ☐ Dried chiles
- ☐ Garlic
- ☐ Lettuce
- ☐ Microgreens
- ☐ Mushrooms (cultivated)
- ☐ Onions
- ☐ Parsnips
- ☐ Potatoes
- ☐ Rutabaga
- ☐ Shallots
- ☐ Spinach
- ☐ Sprouts
- ☐ Sweet potatoes
- ☐ Tomatoes
- ☐ Turnips
- ☐ Winter radishes

FRUITS

- ☐ Apples
- ☐ Frozen fruits
- ☐ Pears

MEATS & FISH

- ☐ Beef
- ☐ Bison
- ☐ Chicken
- ☐ Emu
- ☐ Lamb
- ☐ Panfish
- ☐ Pork
- ☐ Rainbow trout
- ☐ Sausages
- ☐ Smoked fish
- ☐ Turkey
- ☐ Venison
- ☐ Whitefish

EGGS & DAIRY

- ☐ Butter
- ☐ Buttermilk
- ☐ Cheese (cow, goat & sheep milk)
- ☐ Cottage cheese
- ☐ Cream
- ☐ Curds
- ☐ Dips
- ☐ Eggs
- ☐ Ice cream
- ☐ Milk
- ☐ Yogurt

OTHER

- ☐ Baked goods
- ☐ Beer
- ☐ Breads
- ☐ Cider
- ☐ Flours
- ☐ Herbs
- ☐ Hickory nuts
- ☐ Honey
- ☐ Horseradish
- ☐ Jams
- ☐ Maple syrup
- ☐ Pesto
- ☐ Pickles & preserves
- ☐ Popcorn
- ☐ Prepared foods
- ☐ Salsa
- ☐ Sauces
- ☐ Spirits
- ☐ Sunflower oil
- ☐ Tortillas
- ☐ Vinegars
- ☐ Wild rice

Cheeses of the Month: Edam and Gouda

Profile: Edam and Gouda are named after the Dutch towns in which they were invented hundreds of years ago. Shaped into balls or spheres with slightly flattened tops and bottoms, they are often coated with wax. The two cheeses are similar, but Edam is made of part skim milk, while Gouda generally features whole milk. They're most often produced

from pasteurized cow milk, but goat and sheep milk—in some cases unpasteurized milk—are now being used, too. Both varieties make good cooking cheeses, melting to a soft consistency when heated. Although usually eaten when young and quite mild, Edam and Gouda can also be aged (and sometimes brined) and used as a grating cheese. Both take well to smoking, and to flavor additives; Gouda in particular can come flavored with spices, such as cumin or fenugreek.

Wisconsin-made styles include: Smoked, flavored, aged, raw milk, and "regular" Edam and Gouda.

Taste: Butter, faint caramel or butterscotch, nuts. Gouda is rich and slightly sweet; some say there is a note of coffee; Edam is more delicate and straightforward. Both grow drier, saltier, and earthy as they age. Smoked varieties have woodsy flavor.

Texture/Look: Young varieties are firm but smooth, supple, and creamy on the tongue. Aged varieties deepen in color and become hard and crumbly.

Go-withs: Pears, apples, melon, arugula, winter squash, celeriac, potatoes, roasted red peppers, chiles, smoked meats and fish, maple syrup, toasted nuts, dark breads, bean soup, parsley, mustard, cumin, black pepper.

What to drink: Off-dry Riesling, Pinot Gris, Beaujolais, Chardonnay, Belgian farm ale, spiced ale, apple cider, hard cider. Try porter, stout or bock beer with smoked Gouda.

Classic dishes: Stuffed Edam cheese ball.

Try this: Cut equal amounts of celeriac, carrots, and Gouda into julienne strips. Toss with mustard, cream-thinned mayonnaise, and minced chives. Serve over chilled fresh greens or room-temperature wild rice.

February 2014

Sunday	26
Monday	27
Tuesday	28
Wednesday	29
Thursday	30
Friday	31
Saturday	1

Parsnip Gratin
Servings: 4

Parsnips come late in the growing season and are one of the sweetest of the root vegetables; in fact, their natural sugars increase after the frost. What's more, they store well over the winter, and their sweet, herbaceous flavor is a good match for aged cheeses. Here's a simple but royal treatment that's especially delicious with pork loin or rare roast beef.

    1 pound parsnips
    2 tablespoons butter, divided
    1 cup chicken stock
    salt and pepper
    4 tablespoons heavy cream
    ½ cup shredded aged hard Italian-style cheese (Pecorino, Asiago, etc.)

Heat oven to 400 degrees. Peel parsnips with potato peeler; slice into thin rounds or cut into matchsticks. Heat all but about 2 teaspoons of the butter in a skillet over a medium flame. Add parsnips and cook, tossing often, until partially tender, about 3 minutes. Pour in the stock and bring to strong simmer. Lower heat, cover skillet, and gently simmer parsnips until tender, 5–10 minutes. Uncover, raise heat to high, and cook until liquid reduces to a syrupy glaze. Season to taste with salt and pepper. (If you've used salted chicken stock, you may not need additional salt.)

Use the rest of the butter to grease a wide, shallow, medium-sized baking dish. Place half the parsnips in dish. Drizzle with half the cream and sprinkle with half the cheese. Repeat layers with remaining ingredients. Bake until golden brown, about 20 minutes. Serve immediately.

February 2014

Sunday	2
Groundhog Day	
Monday	3
Tuesday	4
Wednesday	5
Thursday	6
Friday	7
Saturday	8

Cumin Gouda Gougères (aka Cheese Puffs)

Servings: 8 (25–30 small puffs total)

Holland's Family Cheese makes a remarkable cumin-flecked cheese, one of numerous varieties in their prize-winning Marieke Gouda line (which includes the cheese that was named the nation's best at the 2013 U.S. Cheese Championship Contest; see page 36 for more info). Their cumin Gouda was the inspiration for this recipe, but feel free to play with the cheese and seasonings to get a completely different flavor profile. Try Cheddar and chili powder for a Southwestern flavor, for example, or Gruyère and minced fresh thyme to go European.

> 7 tablespoons chilled unsalted butter, cut into chunks
> ¾ teaspoon salt
> 1⅓ cups flour
> 4 large eggs
> 1–1½ cups shredded cumin Gouda (such as Marieke Gouda from
>   Holland's Family Cheese)
> ¼ teaspoon cayenne pepper

Heat oven to 400 degrees. Line two baking sheets with parchment paper. Combine cold butter chunks, salt, and 1 cup water in a medium-large saucepan. Bring to boil. Continue to boil until butter is fully melted, about 30 seconds. Remove the pan from the heat. Add the flour all at once and then beat with a wooden spoon until the mixture forms into a smooth ball that looks dry and pulls away from the sides of the pan (this will take only another 30-60 seconds).

Let mixture cool 2–3 minutes and then beat in the eggs, one at a time, making sure each egg is fully incorporated before adding the next. Beat in cheese and cayenne.

Drop by heaping tablespoonfuls onto baking sheets, shaping each one into a rough ball and keeping them at least one inch apart. Bake until puffed, dry-looking, golden-brown all over, and brown-tipped in spots, 30–35 minutes. Serve warm.

February 2014

Sunday	9
Monday	10
Tuesday	11
Wednesday	12
Thursday	13
Friday	14
Valentine's Day	
Saturday	15

Grilled Provolone and Pepperoni Sandwich (And Other Grilled Cheese Ideas)

Servings: 4

Wisconsin has a legacy of sausage-making that stems from its long history of immigration. Try to name all the varieties of sausages made in the state and you'll begin to get an idea of just how many ethnicities have flavored the cooking here. The same thing goes for cheese, too, of course.

Adding to that delicious legacy are the artisan sausage makers and cheesemakers of recent years. Today's nose-to-tail butchers are churning out specialties like Spanish-style chorizo and 'nduja (a spicy spreadable salami), while professional cheeseheads give us ash-covered chèvre and washed rind tomme. The new and re-introduced varieties seem endless, and while it may seem impossible to keep up, it's fun to try.

Here's a simple grilled sandwich with an Italian bent. Use it as a starting place to try your own combinations from the amazing array of cheese and sausage in our state.

> 8 slices Italian-style bread
> olive oil
> 1 cup pizza sauce
> dried oregano
> 6–8 ounces Provolone, sliced
> 2–3 ounces thin-sliced pepperoni
> salt and black pepper to taste

Brush one side of each of the bread slices with olive oil. Spread pizza sauce on other side of each of the bread slices. Lightly sprinkle dried oregano on pizza sauce. Make a sandwich, oiled sides out, with provolone and pepperoni.

Heat a large, heavy skillet or cast iron griddle over medium flame 5–10 minutes. When the pan is hot, grill the sandwiches until both sides are browned and cheese is melted.

Note: Here are some additional combinations to try:

- Cheddar + summer sausage + mustard + rye bread
- Monterey Jack + chorizo + salsa + corn tortillas
- Brick + liver sausage + pickle relish + potato bread
- Chèvre + saucisson + tapenade + French bread

February 2014

Sunday	16
Monday	17
Presidents' Day	
Tuesday	18
Wednesday	19
Thursday	20
Friday	21
Saturday	22

Marieke Penterman, Owner and Licensed Cheesemaker

Holland's Family Cheese
N13851 Gorman Avenue, Thorp
715-669-5230
hollandsfamilycheese.com

Marieke Penterman is a gift to Wisconsin from the country of Holland. She grew up on her parent's 60-cow dairy farm and developed a love of the dairy industry that followed her to college. After receiving a bachelors degree in dairy business, Marieke began a career as a farm inspector. But America called. In 2002 her future husband, Rolf Penterman, emigrated to Thorp, Wisconsin, with his brother and started a 350-cow dairy farm. A year later Marieke married Rolf and joined him in Wisconsin. Once in the United States, Marieke began to hanker for the cheese from home, motivating her to become a licensed cheesemaker. After apprenticing with U.S. and Dutch cheesemakers, Marieke formed the Holland's Family Cheese company to make Gouda cheeses. Just four months after the first batch of Gouda was made, Holland's Family Cheese won Best of Class in the Open Class for flavored semi-soft cheeses at the 2007 U.S. Championship Cheese Contest. Many more awards followed. In 2011 Marieke became the first woman to capture the Grand Master Cheesemaker title at the Wisconsin State Fair. And in 2013 her Marieke Mature Gouda was crowned the U.S. Championship Cheese Contest's top cheese!

Please describe your cheeses. Everything is made at the farm. There is a pipeline straight from the milking room to the cheese vat, and whole cow's milk is turned into cheese within five hours of milking. The freshness of the milk gives the cheese its incredible full flavor. Marieke Gouda is a raw milk product aged for a minimum of 60 days. At that young age, the cheese is very smooth; the flavor becomes more complex and a little sharper as it ages. The Gouda wheels weigh 18 to 20 pounds and are covered with a breathable coating that allows the cheese to continue to age. Plain Marieke Gouda is available aged two-plus years. Flavored Gouda cheese varieties are also available: Cumin, Burning Melange, Clove, Pesto Basil, Black Pepper Mix, and the popular Foenegreek, flavored with foenegreek seeds, which give the cheese a slightly sweet, nutty flavor.

What is it about cheese? Just why is it so wonderful? I love cheese. I think that good food, good drinks, and good company create the best memories and keep the therapist away. Also, I never knew how much work cheesemaking was until I started it. Before, I would pick up a piece of cheese and think, "Wow, that is a lot of money." Now I pick up a piece of cheese and think, "Wow, that cheesemaker doesn't make enough money."

Anything else you'd like to add? Cheesemaking is TEAM work and I'm very fortunate that I have an incredible team that is dedicated and willing to go the extra mile. Without them, we wouldn't be where we are today.

February 2014

Sunday	**23**
Monday	**24**
Tuesday	**25**
Wednesday	**26**
Thursday	**27**
Friday	**28**
Saturday	1

Inspired cheese and beer pairings. ©Wisconsin Milk Marketing Board, Inc. (Digitally altered)

Willi Lehner, owner-operator of Bleu Mont Dairy in Blue Mounds, gained national recognition when the New York Times *included him in a roundup of Dairyland cheesemakers, calling him "the off-the-grid rock star of the Wisconsin artisanal cheese movement." A fixture at the Dane County Farmers' Market since 1988 (and one of a few cheesemakers who vends there year-round), Willi says, "Our specialty cheeses, often made from raw milk, are surface-cured in our underground, simulated cave environment [and] aged from two months to three years, depending on the type of cheese and method of affinage. Our signature cave-aged cheese is bandaged (cloth-bound) Cheddar, made in the tradition of the first Cheddars from the British Isles. The cave is also home to a variety of washed-rind cheeses, including a soft, delightfully stinky gem we call Earth Schmier, inspired by our homestead terroir."*

VEGETABLES (INCLUDES HYDROPONIC & STORED)

- [ ] Arugula
- [ ] Beets
- [ ] Burdock root
- [ ] Carrots
- [ ] Celeriac
- [ ] Dried beans
- [ ] Dried chiles
- [ ] Garlic
- [ ] Lettuce
- [ ] Microgreens
- [ ] Mushrooms (cultivated)
- [ ] Onions
- [ ] Potatoes
- [ ] Rutabaga
- [ ] Shallots
- [ ] Spinach
- [ ] Sprouts
- [ ] Sweet potatoes
- [ ] Tomatoes
- [ ] Turnips
- [ ] Winter radishes

FRUITS

- [ ] Frozen fruits

MEATS & FISH

- [ ] Beef
- [ ] Bison
- [ ] Chicken
- [ ] Emu
- [ ] Lamb
- [ ] Pork
- [ ] Rainbow trout
- [ ] Sausages
- [ ] Smoked fish
- [ ] Turkey
- [ ] Venison
- [ ] Whitefish

EGGS & DAIRY

- [ ] Butter
- [ ] Buttermilk
- [ ] Cheese (cow, goat & sheep milk)
- [ ] Cottage cheese
- [ ] Cream
- [ ] Curds
- [ ] Dips
- [ ] Eggs
- [ ] Ice cream
- [ ] Milk
- [ ] Yogurt

OTHER

- [ ] Baked goods
- [ ] Beer
- [ ] Breads
- [ ] Cider
- [ ] Flours
- [ ] Honey
- [ ] Horseradish
- [ ] Jams
- [ ] Maple syrup
- [ ] Pesto
- [ ] Pickles & preserves
- [ ] Popcorn
- [ ] Prepared foods
- [ ] Salsa
- [ ] Sauces
- [ ] Spirits
- [ ] Sunflower oil
- [ ] Tortillas
- [ ] Vinegars
- [ ] Wild rice

Cheese of the Month: Cheddar

Profile: Entire books can—and have—been written about Cheddar, the most widely consumed cheese on the planet. Developed in a medieval England village of the same name, Cheddar is no longer primarily associated with its town of origin, but with a specific production procedure. Cheddaring is the process in which cheese curds are drained, cut into slabs, piled four high, and then turned repeatedly to compact evenly and smoothly. All cheddars start out white; the orange varieties get their color from a dye made from annatto seeds, which come from the tropical achiote tree. Wisconsin cheesemakers produce more Cheddar than any other state in the Union.

Styles include: White, aged, raw milk, bandaged, blue-veined, mammoth, cave-aged, flavored, smoked.

Taste: The flavor of Cheddar, depending on production method and age, can range from "one note" simple to quite complex, and from thin or very mild to full and extra-sharp. Look for nuts, sweetness, fruit, tang, acidity, grass, spice, caramel.

Texture/Look: Usually smooth, firm, and tightly textured. Aged, artisan varieties can be flaky or crumbly, while most factory-produced Cheddar has a waxy look. Color ranges from creamy white to burnt orange.

Go-withs: Apples, pears, cauliflower, broccoli, onions, chiles, potatoes, sweet corn, ham, bacon, summer sausage, beef, walnuts, salsa, mustard, beer.

What to drink: Young Cheddars pair well with Cabernet Franc, Merlot, Sauvignon Blanc, apple cider, milk. Aged Cheddar goes with Riesling, merlot, Cabernet Sauvignon, Shiraz, Zinfandel, Sauternes, vintage port, sherry, British ale, India pale ale, Belgian-style saison, farmhouse hard cider.

Classic dishes: Beer cheese soup, grilled cheese sandwich, macaroni and cheese, burger with Cheddar, apple pie with Cheddar.

Try this: Cook a pot of soupy black beans flavored with garlic and cumin. Bake your favorite cornbread. Slice sweet onions as thinly as you can. Shred some good extra-sharp Cheddar. Now ladle some beans into a soup plate, crumble the cornbread over the beans. Top with onions and Cheddar. Plow right in.

Sunday	23
Monday	24
Tuesday	25
Wednesday	26
Thursday	27
Friday	28
Saturday	1

New Moon

White Cheddar Chive Soup with Popcorn Garnish

Servings: 8

You've had Cheddar on popcorn, so why not try popcorn on cheddar soup? Like other crispy soup toppers, it adds crunch and contrast, and in this case also makes a great complement to the flavors in the soup.

The mixed colors of the sweet peppers are really pretty in this soup. However, if you're not a fan of sweet peppers, you can substitute broccoli or another vegetable, or replace it with cooked chicken or ham. Or leave the extras out altogether.

4 tablespoons butter
⅓ cup finely chopped shallots
⅓ cup diced sweet orange pepper
⅓ cup diced sweet red pepper
⅓ cup diced sweet yellow pepper
4 tablespoons flour
3 cups milk, heated
3 cups mashed potatoes
3–3 ½ cups chicken or vegetable stock
3 cups grated sharp Cheddar
salt and pepper to taste
4 tablespoons chopped fresh chives
1 ½ – 2 cups popped popcorn

Heat butter in heavy soup pot over a medium flame. Add shallots and cook, stirring occasionally, for 2–3 minutes. Add diced sweet peppers; cook over medium-low heat until vegetables are tender, 6–8 minutes.

Stir in flour and cook gently 2 minutes, stirring often. Whisk in hot milk, potatoes, and chicken stock until smooth. Simmer 10–15 minutes.

Remove from heat, let stand a few moments to cool down slightly, then whisk in cheese a little at a time. Season soup with salt and pepper to taste. Stir in chives. Reheat gently.

Serve with popped popcorn scattered over the top of each bowl.

March 2014

Sunday 2

Monday 3

Tuesday 4

Wednesday 5

Ash Wednesday

Thursday 6

Friday 7

Saturday 8

Sour Cream and Cheddar Enchiladas with Chipotle Gravy
Servings: 6

We try to kid ourselves in March. "This is the last of the snow," we claim with bravado. But then we're back in the garage again, pulling out the shovel for one more round with the white stuff. If we're lucky, there's a dish like this one waiting in the oven when we come in from the cold… this one last time.

  2 ½ cups chicken or vegetable stock, divided
  3 chipotles (dried jalapeños), stemmed and seeded
  3 tablespoons chili powder
  2 tablespoons vegetable oil, plus a little for the baking dish
  ¾ cup finely chopped onion
  2 teaspoons minced garlic
  1 teaspoon oregano
  1 teaspoon salt (omit this if using canned or boxed stock)
  1 tablespoon flour
  1 pound shredded mild Cheddar
  12 ounces sour cream
  ½ cup chopped green onions
  1 teaspoon ground cumin
  2 tablespoons chopped cilantro
  12–14 corn tortillas, at room temperature
  Garnishes: ½ cup shredded Cheddar and 2 tablespoons
    minced green onion

To make gravy, heat 1 cup of the stock and pour it over chipotles; let soften 10–15 minutes. Add chili powder. Purée the mixture. Heat 2 tablespoons oil in a saucepan over a low flame. Add onions, garlic, and oregano; cook until onions are tender, stirring occasionally. Stir in 1 cup stock, chipotle purée, and salt (if using); simmer 15–20 minutes. Combine flour with the remaining stock; whisk into sauce until thickened.

Combine Cheddar, sour cream, green onions, cumin, and cilantro in a bowl. Heat a cast iron skillet over a medium flame. Heat oven to 350 degrees. Oil a large baking dish. Briefly heat a tortilla on each side in skillet to soften it. Dip it into the gravy and place it on a plate. Spread 3–4 tablespoons filling across the middle, roll it up and place it in the baking dish. Continue this process, packing the filled enchiladas into the dish. Pour remaining gravy over and around enchiladas. Bake until bubbly, 25–35 minutes. Sprinkle with garnishes; let stand 10 minutes before serving.

March 2014

Sunday	9
Daylight Saving Time Begins	
Monday	10
New Moon	
Tuesday	11
Wednesday	12
Thursday	13
Friday	14
Saturday	15

Peruvian Potatoes with Hard-Cooked Eggs and Queso Fresco

Servings: 6

Queso Fresco is a semi-firm fresh white cheese that has a mild, lightly salty, and tangy taste. When heated, the cheese softens but doesn't melt, giving it a lovely mouth-feel. The most popular cheese in Mexico, Queso Fresco is typically crumbled and sprinkled onto enchiladas, tamales, soups, salads, and many more preparations. Here we've featured it in a Peruvian-style dish that's a little like scalloped potatoes, only spicier and more stew-like, and with the added pleasure of hard-cooked eggs. It makes a hearty, meatless entrée or can be served as a side dish.

2 pounds red potatoes, scrubbed
6 eggs
2 tablespoons butter
1 cup finely chopped red onion
2–3 jalapeños, seeded and minced
1 tablespoon minced garlic
1 ½ cups milk
4–6 ounces Queso Fresco, crumbled
2 tablespoons chopped cilantro
salt and pepper to taste

Boil potatoes in salted water until nearly tender. Meanwhile, place eggs in a saucepan and cover with cold water. Place over a medium flame. When the water comes to a simmer, turn off the heat, cover the pan, and let the eggs stand 8 minutes. Drain and immerse eggs in ice water to cool them. Peel and cut them into sixths.

Melt butter in a large, deep skillet over medium flame. Add red onions; cook 3–4 minutes, stirring occasionally. Add jalapeños and garlic; cook another 3 minutes or so.

Drain the cooked potatoes; when they've cooled off a bit, cut them into chunks or thick slices. Gently stir potatoes and milk into the cooked onions. Simmer until potatoes absorb some of the liquid. Reduce heat to low, stir in cheese and eggs, and heat through.

Stir in cilantro plus salt and pepper to taste. Serve in shallow bowls.

March 2014

Sunday	16
Monday	17
St. Patrick's Day	
Tuesday	18
Wednesday	19
Thursday	20
First Day of Spring	
Friday	21
Saturday	22

Cottage Cheese Honey Pie with Dried Cherries

Servings: 8

This is not fancy food. This is homey food. It's simple and satisfying. It's sweet enough for dessert but not so sweet that you couldn't serve it for brunch. If you want to refine it a little, purée the cottage cheese in a food processor or blender before adding it to the filling mixture. It will be smoother, but it will still be homey.

Note that this recipe calls for heating a baking sheet and placing the pie on the hot sheet to bake it. This provides a heat boost that helps set the crust and prevents it from getting soggy.

3 extra-large eggs or 4 large eggs
⅓ cup honey
1 container (16 ounces) cottage cheese
⅔ cup dried cherries (coarsely chopped if they are large)
½ teaspoon vanilla extract
½ teaspoon salt
1 unbaked deep-dish 9-inch pie crust
ground cinnamon

Place a foil-lined baking sheet in the oven and heat oven to 375 degrees.

Beat the eggs and honey in a large bowl until smooth. Add cottage cheese, dried cherries, vanilla extract, and salt. Mix well. Pour the mixture into the pie crust. Stir the filling in the crust to evenly distribute the dried fruit. Sprinkle cinnamon lightly over entire surface.

Place pie on hot baking sheet. Bake until filling is barely set in the middle, about 1 hour. Cool the pie completely on a wire rack. Serve it at room temperature.

March 2014

Sunday	**23**
Monday	**24**
Tuesday	**25**
Wednesday	**26**
Thursday	**27**
Friday	**28**
Saturday	**29**

Kerry Henning, Master Cheesemaker and President

Henning's Cheese
20201 Point Creek Road, Kiel
920-894-3032
henningscheese.com

Kerry Henning, certified as a Master Cheesemaker for Cheddar, Colby, and Monterey Jack cheese varieties, says his cheese factory stayed true to making traditional wheels of cheese the same way his grandfather and father did, even though the once almost universal way of making cheese became a thing of the past with the introduction of automation to reduce costs. In the last five years, however, there has been a resurgence of sales for the traditional wheels, prompting Kerry to say, with a slight smile, "If you stay backwards long enough, you'll become in vogue again." And wheel cheeses are more flavorful than mass-produced forty-pound blocks, according to Henning.

Since Henning's Cheese is a relatively small company, it is able to capitalize on the trend of flavored cheeses. Kerry can work with his customers to craft a flavored cheese that fits a specific market and not have to worry about high volume production. The factory is also the only manufacturer in the nation that is still making mammoth wheels of Cheddar. They are big attention-getters for grocery store openings and other celebration events, especially when the wheels are artfully hand-carved. The gigantic wheels can weigh as much as 12,000 pounds—or 5.5 tons. That's a huge hunk of cheese!

What types of cheeses do you make besides Cheddar? We manufacture Colby, Monterey Jack, Colby Jack, regular-size string cheese and small, skinny pieces of string cheese that we market as "Mozza Whips."

Is there a fourth generation of Henning cheesemakers in the wings? I now have two nieces in sales and I'm training my son and nephew to be the next generation of cheesemakers at Henning's.

Why do you think Wisconsin cheese is the best in the world? One thing I must say is that great cheeses are made all over the world. The best cheeses are made with passion. You must appreciate what you do and want only the best for your customers. Wisconsin has great soils and climate for making cheese. We have a great infrastructure as well—schools, roads, support businesses—and a good work ethic. We take a lot of this for granted, but it's what helps make Wisconsin the best state to make cheese in.

©Wisconsin Milk Marketing Board, Inc.

March 2014

	Sunday	30
51		
	New Moon	

	Monday	31

	Tuesday	1

	Wednesday	2

	Thursday	3

	Friday	4

	Saturday	5

Ode to spring. Photo by Terese Allen.

What's your favorite cheese in springtime? Here's what several specialists chose:

VICKI THINGVOLD, MEISTER CHEESE COMPANY:
"***Wild Morel & Leek Jack***—[it has] an earthy springtime flavor."

PAULA HOMAN, RED BARN FAMILY FARMS:
"***Red Barn Edun***—a New Zealand-style raw milk Cheddar—because it tastes clean, fresh, and pure. Our customers say it has a 'happy flavor.'"

ANDY HATCH, UPLANDS CHEESE:
"Fresh ***chèvre***, because it tastes like spring."

FELIX THALHAMMER, CAPRI CHEESE:
"***Capri Fromage Blanc***, because it is light and so tasty."

VEGETABLES (INCLUDES HYDROPONIC & STORED)

- [ ] Burdock root
- [ ] Celeriac
- [ ] Cucumbers
- [ ] Dried beans
- [ ] Dried chiles
- [ ] Garlic
- [ ] Jerusalem artichokes
- [ ] Microgreens
- [ ] Mushrooms (cultivated)
- [ ] Potatoes
- [ ] Radishes
- [ ] Ramps
- [ ] Shallots
- [ ] Sprouts
- [ ] Tomatoes

HERBS & FRESH GREENS

- [ ] Arugula
- [ ] Chives
- [ ] Dandelion greens
- [ ] Lettuces
- [ ] Microgreens
- [ ] Nettles
- [ ] Parsley
- [ ] Purslane
- [ ] Sorrel
- [ ] Spinach
- [ ] Watercress

MEATS & FISH

- [ ] Beef
- [ ] Bison
- [ ] Chicken
- [ ] Emu
- [ ] Lamb
- [ ] Pork
- [ ] Rabbit
- [ ] Rainbow trout
- [ ] Sausages
- [ ] Smoked fish
- [ ] Turkey
- [ ] Venison
- [ ] Whitefish

EGGS & DAIRY

- [ ] Butter
- [ ] Buttermilk
- [ ] Cheese (cow, goat & sheep milk)
- [ ] Cottage cheese
- [ ] Cream
- [ ] Curds
- [ ] Dips
- [ ] Eggs
- [ ] Ice cream
- [ ] Milk
- [ ] Yogurt

OTHER

- [ ] Baked goods
- [ ] Beer
- [ ] Breads
- [ ] Cider
- [ ] Flours
- [ ] Ginseng
- [ ] Hickory nuts
- [ ] Honey
- [ ] Horseradish
- [ ] Jams
- [ ] Maple syrup
- [ ] Pesto
- [ ] Pickles & preserves
- [ ] Popcorn
- [ ] Prepared foods
- [ ] Salsa
- [ ] Sauces
- [ ] Spirits
- [ ] Sunflower oil
- [ ] Tortillas
- [ ] Vinegars
- [ ] Wild rice

Cheese of the Month: Chèvre

Profile: Chèvre, which means goat in French, refers to fresh, white, unripened goat cheese that is packed into small containers or rolled into logs, disks, pyramids, and other shapes. Cheesemakers often coat chèvre with herbs, nuts, or ash, and sometimes they ripen and age the cheeses too. Chèvre is an ancient cheese; it was introduced to the United States by Laura Chenel and its Midwestern pioneers were Anne Topham and Judy Borree of Fantôme Farm in Ridgeway, Wisconsin, who launched their business in the 1980s. Until 2012, when cheesemaker Topham retired from commercial production, the farm's fresh chèvre was a beloved mainstay at the Dane County Farmers' Market in Madison. Milk production of goats is seasonal—usually from mid-March through October. That means supplies are limited…so get it while you can.

Wisconsin brands include: LaClare Farm, Montchevré-Betin, Dreamfarm.

Taste: Tang, light acidity, faint barnyard.

Texture: Delicate, smooth, moist, very creamy, and spreadable. Some tubbed varieties have an almost whipped texture. Aged varieties can become crumbly.

Go-withs: Strawberries, raspberries, hardy kiwi (a small fruit resembling kiwifruit that can grow in Wisconsin), currants, dried fruits, beets, fresh spinach, tomatoes, mushrooms, sweet corn, garlic, toasted nuts, pepper, herbs, smoked fish, prosciutto, chutney, crostini.

What to drink: Sauvignon Blanc, Chenin Blanc, Pinot Grigio, light Rieslings, Gewürztraminer, Pinot Noir, New Glarus Brewing Company's Wisconsin Belgian Red or Raspberry Tart, India pale ale, wheat beer, apple or pear cider.

Classic dishes: Crostini, beet and chèvre salad, mushroom and chèvre tart.

Try this: From Diana Kalscheur Murphy of Dreamfarm in Cross Plains: "We love cheese-baked eggs: Butter a muffin tin, place a big tablespoon of cheese in the bottom of each and flatten. Into each cup crack an egg, add 1 teaspoon cream, salt and pepper, herbs if you like, and top with a grating of your favorite cheese. Place muffin tin into a larger (9-by-13-inch) pan and add water to about ¾-inch deep. Bake in preheated oven at 350 degrees for about 25 minutes. Enjoy as is, or place between good chewy bread or toast for a sandwich."

April 2014

Sunday 30

Monday 31

Tuesday 1

April Fools' Day

Wednesday 2

Thursday 3

Friday 4

Saturday 5

Polenta with Blue Cheese and Caramelized Onions
Servings: 6–8

A steaming bowl of polenta napped with slow-cooked onions and sweet-salty blue cheese gives a warm glow to early spring meals. Part of polenta's appeal is its creaminess, and to get it that way, most recipes require prolonged stirring in a pot on the stove. But there's a much easier way, a no-stir method in which you simply combine cornmeal and a flavoring liquid in a baking dish and then let it bake its way to smooth perfection. We just love it.

Here's what else we love: how wonderfully polenta goes with cheese (and seasons) of all kinds. Try it with feta and grilled sweet red peppers in summer, Parmesan and puréed squash in fall, and Gruyère and sautéed mushrooms in winter.

> olive oil or softened butter
> 1 cup yellow medium-grind cornmeal
> 2 cups milk
> salt and pepper
> 3 tablespoons unsalted butter, divided
> 1 teaspoon sugar
> 4 ounces crumbled blue cheese

Heat oven to 350 degrees. Oil or butter a large (approximately 2-quart) enamel casserole dish or ovenproof nonstick skillet. Add cornmeal, milk, 2 cups water, 1 teaspoon salt, and some grindings of pepper, and stir the mixture for a moment. Don't worry that the mixture doesn't come together; it will do so as it cooks. Bake uncovered 40–45 minutes.

Meanwhile, melt 3 tablespoons butter in a large, heavy skillet over medium-low flame. Stir in onions. Cook them slowly, stirring often, until they are limp and golden, 25–30 minutes. Adjust the heat as needed to prevent the onions from cooking too quickly or scorching. Sprinkle the sugar plus salt and pepper to taste over the onions. Cook the onions another 5–10 minutes, until they are golden-brown. Keep warm.

After the polenta has baked 40–45 minutes, take a look; if it hasn't absorbed nearly all the liquid, bake another 5–10 minutes longer. Stir in onions and cheese. Serve the polenta within 5–10 minutes (it will thicken as it cools off).

April 2014

Sunday	6
Monday	7
Tuesday	8
Wednesday	9
Thursday	10
Friday	11
Saturday	12

Warm Mushroom Salad on Spring Greens

Servings: 4

Now that outdoor farmers' markets are re-opening and the garden season has begun (finally!), the fervor for fresh veggies is high. But since most local crops are still in the bedding-plant stage, there aren't a great many vegetable choices just yet. To the rescue come mushroom vendors—the growers who bring us white buttons, criminis, oysters, portobellas, and other cultivated varieties, plus those hard-hunting morel foragers, who have but a brief and much-lauded season to gather these prized morsels, and it is almost here.

With their characteristic meaty savor—now commonly known as umami—mushrooms are a neat solution to an omnivore's "flesh or plants tonight?" dilemma. And mushrooms pair brilliantly with cheese, another umami-packed local product. Here, sautéed 'shrooms are accented with ricotta salata, a firm, aged ricotta cheese.

Dressing:
1 tablespoon good vinegar (use your favorite)
½ teaspoon Dijon-style mustard
2–3 tablespoons extra virgin olive oil
salt and pepper
Salad:
2–3 tablespoons extra virgin olive oil
¼ cup chopped spring garlic shoots or ramps
1 pound fresh mushrooms, sliced
¼ cup dry white wine
1–1 ½ cups spring herbs (parsley, chives, sorrel, cilantro, etc.),
  chopped
salt and pepper
4–6 cups mesclun (assorted young salad leaves)
½ cup crumbled or diced ricotta salata

Combine vinegar and mustard. Whisk in oil a little at a time. Add salt and pepper as needed. Set dressing aside. Heat oil in one large or two medium skillets over medium flame. Add garlic shoots or ramps and cook 10–20 seconds. Raise heat to medium-high, add mushrooms, and cook, stirring often, until tender. Stir wine into pan and let liquid reduce to a glaze. Let stand 5 minutes to cool slightly, then stir in herbs plus salt and pepper as needed.

Toss greens with just enough dressing to lightly coat them. Mound on serving plates, top with mushrooms, and sprinkle with cheese.

April 2014

Sunday 13

Monday 14

Passover Begins (Sunset)

Tuesday 15

Tax Day

Wednesday 16

Thursday 17

Friday 18

Good Friday

Saturday 19

Smoked Trout Chèvre Cheesecake

Servings: 8–10

This unusual make-ahead preparation takes cheesecake from the dessert course to the main course. Pair it with a simple spring salad—watercress tossed with good oil and vinegar would be ideal—for a lunch or supper dish that will wow your guests. The recipe was shared by Jill L. K. McLeod of Middleton.

Crust:
1 ½ cups broken-up pretzels
6 tablespoons butter, melted
2 egg whites
Filling:
12 ounces cream cheese, softened
6 ounces chèvre (soft goat cheese)
3 whole eggs
2 egg yolks
¼ cup freshly squeezed lemon juice
1 teaspoon ground mace
1 tablespoon sugar
salt and ground white pepper to taste
1 cup grated apple
¼ – ⅓ pound smoked rainbow trout, boned and flaked

Prepare the crust: Heat oven to 350 degrees. Crush pretzels in a food processor, or put pretzels in a zippered plastic bag and crush them with a rolling pin. Combine pretzel crumbs, melted butter, and egg whites in a bowl. Pat mixture into the bottom of an 8-inch spring form pan, and bake 10 minutes. Leave the oven on after removing the crust.

Prepare the filling: Beat the cream cheese and chèvre in a bowl with electric beaters until smooth. Add the whole eggs, beating well after each addition. Add the yolks, lemon juice, mace, sugar, salt, and pepper; mix on low speed until blended. Fold in the grated apple and flaked trout with a rubber spatula. Turn the filling onto the spring form pan. Bake until cheesecake is set and starting to brown on top, about 55 minutes. Cool to room temperature, then cover and chill for 3 hours or longer before serving.

April 2014

	Sunday	20
	Easter	

	Monday	21

	Tuesday	22
	Passover Ends	
	Earth Day	

	Wednesday	23

	Thursday	24

	Friday	25
	Arbor Day	

	Saturday	26

Scott Erickson, Co-owner and Master Cheesemaker

Bass Lake Cheese Factory
598 Valley View Trail, Somerset
715-247-5586
blcheese.com

Scott Erickson began making cheese in 1984 at the Bass Lake Cheese Factory in Somerset, a factory which was established in 1918 and had changed ownership and names several times before he became a cheesemaker there. He fell in love with the hands-on process and the creativity this allowed, initially making cow milk cheeses such as Cheddar, Colby, and Jack, and then goat and sheep milk cheeses. Seven years later Scott and his wife Julie bought the factory and began specializing in artisan cheeses. Scott is now a six-time Master Cheesemaker, and is certified in Colby, Cheddar, Jack, Muenster, chèvre, and Juustoleipa.

Tell us what changes new ownership brought to the company. We redirected the path Bass Lake Cheese would take in becoming a destination spot for small-batch specialty cheeses, with a focus on educating the consumer and direct consumer sales. A full liquor license, full kitchen, and indoor and outdoor seating areas were added to promote Bass Lake cheeses. We feature our cheeses at wine and cheese samplings held once a month and in dishes made in our kitchen: pizzas, grilled cheese sandwiches, sub sandwiches, homemade soups, and deep-fried Juusto. Future plans include a rooftop garden and deck for our faithful customers to enjoy.

What's your favorite thing about cheese? The magic it provides in the make and the comfort it provides in the eat!

Where do you get your inspiration to craft a new cheese? I collect vintage cheesemaking books describing earlier ways Old World cheeses were made. I have also traveled to the eastern European countries of Romania, Bulgaria, Macedonia, and Albania to help people establish privatized cheese plants. In turn, I was exposed to century-old cheeses they produced and in some cases, learned their cheesemaking processes and techniques. I have also learned about making many other cheeses by attending Master Cheesemaker courses at the Center for Dairy Research in Madison at the University of Wisconsin.

Which of the cheeses you make is your favorite? My favorite cheese changes as often as the day of the week. Today our four-year Cheddar is my favorite; yesterday my Juustoleipa (deep-fried) was my favorite; and Sunday my chèvre on fresh baked pizza filled my cheese desire.

What is it about cheese that makes it an ideal ingredient to cook with? Each cheese has its own characteristics and flavor profiles, allowing cooks and chefs unlimited creativity in the foods they make.

April   2014

Sunday	27

REAP's Spring Gala

Monday	28

Tuesday	29

New Moon

Wednesday	30

Thursday	1

Friday	2

Saturday	3

Bruce Workman (Edelweiss Cheese) giving advice to the milk. Photo by Jeanne Carpenter.

If there's any month when simple is best, it's got to be May. When we asked cheese-ophiles to tell us springtime dishes they love to eat with their crafted specialties, it seemed that the less complicated the combination, the more mouth-watering it sounded. Willi Lehner of Bleu Mont Dairy, for instance, says he goes for his Bandaged Cheddar grated over steamed asparagus. Joe Burns of Brunkow Cheese likes his French-style tomme, called Little Darling, with fresh greens and a dry Rosé. And BelGioioso's marketing vice-president, Francis Wall, thinks the company's mascarpone is delicious with fresh berries.

VEGETABLES

- [ ] Asparagus
- [ ] Bok choy
- [ ] Broccoli
- [ ] Dried beans
- [ ] Dried chiles
- [ ] Fiddlehead ferns
- [ ] Green garlic
- [ ] Green onions
- [ ] Morels (wild)
- [ ] Mushrooms (cultivated)
- [ ] Peas
- [ ] Radishes
- [ ] Ramps
- [ ] Salad turnips
- [ ] Sprouts

HYDROPONIC & WINTER-STORED VEGETABLES

- [ ] Burdock root
- [ ] Cucumbers
- [ ] Garlic
- [ ] Jerusalem artichokes
- [ ] Potatoes
- [ ] Tomatoes

HERBS & FRESH GREENS

- [ ] Arugula
- [ ] Chard
- [ ] Chives
- [ ] Cilantro
- [ ] Dandelion greens
- [ ] Dill
- [ ] Lettuces
- [ ] Microgreens
- [ ] Mint
- [ ] Nettles
- [ ] Parsley
- [ ] Purslane
- [ ] Sorrel
- [ ] Spinach
- [ ] Watercress

FRUITS

- [ ] Frozen fruits
- [ ] Rhubarb
- [ ] Strawberries

MEATS & FISH

- [ ] Beef
- [ ] Bison
- [ ] Chicken
- [ ] Emu
- [ ] Lamb
- [ ] Pork
- [ ] Rabbit
- [ ] Rainbow trout
- [ ] Sausages
- [ ] Smoked fish
- [ ] Turkey
- [ ] Venison
- [ ] Whitefish

EGGS & DAIRY

- [ ] Butter
- [ ] Buttermilk
- [ ] Cheese (cow, goat & sheep milk)
- [ ] Cottage cheese
- [ ] Cream
- [ ] Curds
- [ ] Dips
- [ ] Eggs
- [ ] Ice cream
- [ ] Milk
- [ ] Yogurt

OTHER

- [ ] Baked goods
- [ ] Beer
- [ ] Breads
- [ ] Cider
- [ ] Edible flowers
- [ ] Flours
- [ ] Ginseng
- [ ] Hickory nuts
- [ ] Honey
- [ ] Horseradish
- [ ] Jams
- [ ] Maple syrup
- [ ] Pesto
- [ ] Pickles & preserves
- [ ] Popcorn
- [ ] Prepared foods
- [ ] Salsa
- [ ] Sauces
- [ ] Spirits
- [ ] Sunflower oil
- [ ] Tortillas
- [ ] Vinegars
- [ ] Wild rice

Cheese of the Month: Blue

Profile: Veined cheeses weren't invented, as such; they happened long ago, when cave-stored cheeses developed mold naturally. Some very brave—or very hungry—soul must have decided to taste, instead of toss, a gnarly looking specimen, and thus  discovered the savory attraction of a cheese "gone bad." (Thank you, whoever you were.) Today, Penicillium strains may be mixed with the milk or curds, injected into the cheese, or sprayed over it, but no matter how they are introduced, they need oxygen for sustenance. To provide it, producers needle the cheese; that is, they pierce it with metal skewers to feed the bacteria and form the veins. Wisconsin produces more than 80 percent of the nation's blue cheeses.

Wisconsin-made styles include: Cow, goat, or sheep milk, or a combination of milks; raw milk blue, aged blue, smoked blue, Italian-style gorgonzola. Top quality Wisconsin brands include Salemville Cheese Cooperative's Amish Blue, LeClare Farm's Ziege Zacke Blue, Organic Valley's Kickapoo Blue, and any veined varieties from Hook's Cheese Company or Swiss Valley Farms.

Taste: Piquancy, pleasant mold, salt, pepper, sweetness, earth. Grows sharp, acidic, and strong with age.

Texture/Look: From creamy and spreadable when young, to firm and crumbly as time passes, and to hard and suitable for grating with long age. All varieties show colored veining in blue-green shades, and color also varies in the main part of the cheese, from ivory to golden-brown.

Go-withs: Pears, apples, asparagus, green beans, cauliflower, beets, potatoes, Romaine, toasted nuts, honey, dried fruits, tarragon, chives, rosemary, bacon, burgers, polenta, pasta.

What to drink: Pinot Noir, late harvest Riesling, Pinot Grigio, sparkling wine, Muscat, port, barley wine, chocolate stout, abbey-style ale, martini, manhattan.

Classic dishes: Blue cheese dressing, cobb salad, buffalo chicken wings, grilled steak with blue cheese.

Try this: From Tony and Julie Hook of Hook's Cheese in Mineral Point: "Slice Hook's Duda Gouda and eat it with strawberry preserves spread on it. Or mix Hook's gorgonzola or any of our blue cheeses with ground beef, form into patties, and roll in pepper to make 'Black and Blue Burgers.'"

Photo ©Wisconsin Milk Marketing Board, Inc.

May 2014

	Sunday	27
Monday	28
Tuesday	29
Wednesday	30
Thursday	1
Friday	2
Saturday	3

Herb-Packed Fresh Cheese Dip

Servings: 4–6

We think the best time to make this spring-inspired dip is right after thinning the seedlings in one's herb garden. Not only are the herbs at their most delicate and vibrant at this point, it's also a great way to feature all of them at once. Serve the dip with a multicolored platter of veggies from the farmers' market or CSA box—asparagus, green onions, radishes, raw peas, etc. Or thin the dip with milk and lemon juice and use it as a dressing for a green salad.

    1 cup ricotta or cottage cheese
    ¼ cup milk or plain yogurt
    3 tablespoons chopped fresh parsley
    3 tablespoons chopped fresh dill
    3 tablespoons chopped fresh cilantro
    2 tablespoons chopped fresh chives or ramps
    2 teaspoons finely grated lemon zest
    2 tablespoons fresh lemon juice
    salt and freshly ground black pepper to taste

Blend ricotta or cottage cheese and milk or yogurt in a food processor or with an immersion blender until as smooth as possible. Stir in parsley, dill, cilantro, chives or ramps, lemon zest, and lemon juice. Season with salt and pepper to taste. Transfer dip to a bowl, cover, and chill until ready to serve.

May 2014

Sunday	4
Monday	5
Cinco de Mayo	
Tuesday	6
Wednesday	7
Thursday	8
Friday	9
Saturday	10

Frittata with Pea Shoots and Feta Cheese

Servings: 4 as a main course; 8 as part of an appetizer spread

Pea shoots may sound like some sort of new-wave gourmet ingredient, but they are merely the trimmings from early pea vines. They have a sweet vegetal flavor and a tender bite.

> 8 large eggs
> 2 tablespoons chopped fresh cilantro or mint leaves
> salt and freshly ground black pepper to taste
> 3 tablespoons butter, divided
> ⅓ cup finely chopped spring onions
> 4 cups chopped pea shoots
> ⅓ cup crumbled feta
> 2 tablespoons chopped kalamata olives (optional)

Beat eggs in a bowl with cilantro or mint, salt, and pepper, and set aside.

Melt 1 ½ tablespoons butter in a 9- or 10-inch nonstick skillet (one with sloping sides) over a medium-high flame. Add onions and cook, stirring often, 1–2 minutes. Add pea shoots; cook, stirring often, until greens wilt, 2–4 minutes. (If there's liquid in the pan at this point, raise heat to high and cook until it has evaporated.)

Reduce heat to medium-low; add another tablespoon butter to pan. Stir in eggs, cover pan, and cook until the eggs are mostly set, 8–10 minutes. (The top will still look a little wet at this point.) Use a spatula to shape and smooth the outer edges of the frittata. Lay a towel on a counter. Slide spatula underneath frittata to loosen it, and then glide it onto a platter. Place platter on the towel.

Add the remaining ½ tablespoon butter to pan. Wearing oven mitts, invert the hot pan over the top of the frittata (if the butter drips, it will go onto the frittata). Grasp the pan-platter construction with both heat-protected hands, and then flip it upside down, setting it back onto the towel. The frittata will now be back in the pan (and on its second side). Sprinkle with feta and return it to the heat to finish cooking a moment or two. To serve, loosen frittata from pan and slide it back onto the platter. Sprinkle with chopped olives, if desired. Serve this hot, warm, or at room temperature.

May 2014

Sunday	11
Mother's Day	
Monday	12
Tuesday	13
Wednesday	14
Thursday	15
Friday	16
Saturday	17

Grilled Asparagus and Gorgonzola Roll-ups

Servings: 10–15

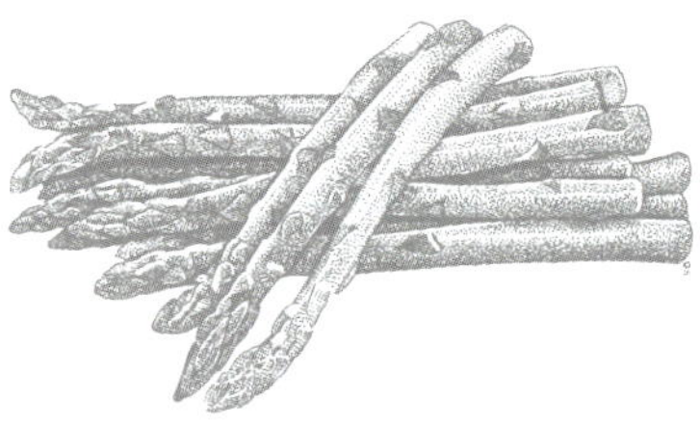

Grilling season can't come soon enough in Wisconsin. We haul out the Webers, gas grills, and backyard fire pits the minute the last snowflake has melted, filling them with brats, burgers, and any other grillables we can get our hands on. By May, many of us are already looking for something more creative—and maybe a little bit lighter—to serve up. Here's an unusual spring appetizer just made for that hungry group crowding around your outdoor hearth.

> 1 package (8 ounces) cream cheese, softened
> 6 ounces gorgonzola, softened
> 2–3 tablespoons minced fresh chives or green garlic shoots
> 1 loaf (1 ½ pounds) soft white or whole wheat sandwich bread, crusts removed
> 20–30 fresh, slender asparagus spears, blanched and blotted dry with towels
> extra-virgin olive oil

Beat cream cheese, gorgonzola, and chives in a bowl until smooth.

Use a rolling pin to flatten and roll out each of the bread slices. Spread a layer of the cheese mixture over each slice. Cut asparagus spears to fit the length of the bread slices. Place 2 spears, tips facing opposite directions, on the long edge of a bread slice and roll it up tightly. Brush it with olive oil. Repeat this until all the filling is used up. At this point the rolls can be tightly wrapped in plastic wrap and refrigerated until you're ready to cook and serve them.

Prepare coals on an outdoor grill (or heat a gas grill) to medium-low heat. The roll-ups may be placed in a grill basket or directly on the grill. Place them about 5 inches from heat and grill until lightly browned, turning once or twice, 3–5 minutes. (Alternatively, the roll-ups may be cooked under a broiler, turning them often). Serve immediately.

May 2014

<table>
<tr><td>Sunday</td><td>18</td></tr>
<tr><td>Monday</td><td>19</td></tr>
<tr><td>Tuesday</td><td>20</td></tr>
<tr><td>Wednesday</td><td>21</td></tr>
<tr><td>Thursday</td><td>22</td></tr>
<tr><td>Friday</td><td>23</td></tr>
<tr><td>Saturday</td><td>24</td></tr>
</table>

Katie Hedrich, Cheesemaker and Marketing Director

LaClare Farms
W2994 County HH, Malone
920-418-2302
laclarefarm.com

Perhaps it was a stroke of luck that Katie Hedrich's folks, Larry and Clara, purchased a farm in 1978 that came with two milking goats, two peacocks, and a flock of laying hens, because this was the spark that ultimately led to commercial dairy goat milk production. Today the herd is 450 strong and milking is a year round operation.

As a child Katie was active in 4H and showed her goats, but it wasn't until she accompanied her parents to Holland in 2009 to tour creameries and goat operations that she became smitten with the art of cheesemaking. Upon her return, she says, she begged the plant manager for a job at the creamery that was then making their cheese—and she got one.

It didn't take long for Katie to garner top honors in competitions. LaClare Farms' first artisan cheese—Evalon, a Gouda-style, raw milk cheese—was named the 2011 U.S. Championship Cheese. At the same time, Katie became the youngest cheesemaker ever (and second woman at the time) to win big in this competition. The family of LaClare goat cheeses has grown to include flavored Evalons (fenugreek for a nutty, maple flavor, or cumin for a savory, spicy flavor), fresh chèvre, Cheddar, and cheese curds.

What's your favorite thing about cheese? I love how simple yet complex cheese is—milk, culture, rennet, and salt. It is so fascinating to me that you start with just a few ingredients and create so many different products!

What features make your cheeses stand out?
The feature that makes our cheeses stand out is the quality of our milk. In 2004 my parents and a few other dairy goat farmers began the Quality Dairy Goat Producers Co-op, which defined what quality goat milk is. The USDA actually adopted the definition as the universal definition. People often tell me they don't like goat cheese, but when they try ours they say, "Oh, that's actually pretty good."

Could you blind-taste your cheese in a lineup of similar cheeses? If our Evalon was in the lineup, absolutely!

Photo by Becca Dilley.

May 2014

Sunday **25**

Monday **26**

Memorial Day

Tuesday **27**

Wednesday **28**

New Moon

Thursday **29**

Friday **30**

Saturday **31**

Diana Murphy (Dreamfarm) and crew. Photo by Becca Dilley.

A day in the life of Farmer John (Farmer John's Cheese). Photo by Bill Lubing.

Is there a difference between grass-fed and conventional cheese? Most definitely, according to a report published by the Wisconsin Department of Agriculture in 2013. During a four-year project, cheese made from pasture-based milk was compared to cheese produced from the milk of confinement farms. The former was more golden in color, creamier in texture, and had more complex flavor and aroma. In professional taste tests, almost all the participants preferred the taste, appearance, mouthfeel, and aroma of cheese made with pasture milk.

Surprised? We weren't either. And we wouldn't be surprised if pasture-based cheese produced during spring tastes the best of all. Chris Roelli of Roelli Cheese, for instance, likes grass-based cheese at this time of year, "for the sweetness of fresh grass."

VEGETABLES

- [ ] Asparagus
- [ ] Baby beets
- [ ] Baby carrots
- [ ] Baby zucchini
- [ ] Bok choy
- [ ] Broccoli
- [ ] Chinese cabbage
- [ ] Dried chiles
- [ ] Garlic scapes
- [ ] Green onions
- [ ] Mushrooms (cultivated)
- [ ] Peas
- [ ] Radishes
- [ ] Salad turnips
- [ ] Spring onions
- [ ] Sprouts

HYDROPONIC & WINTER-STORED VEGETABLES

- [ ] Cucumbers
- [ ] Potatoes
- [ ] Tomatoes

HERBS & FRESH GREENS

- [ ] Arugula
- [ ] Chard
- [ ] Chives
- [ ] Cilantro
- [ ] Collards
- [ ] Dill
- [ ] Kale
- [ ] Kohlrabi
- [ ] Lettuces (leaf and head)
- [ ] Microgreens
- [ ] Mint
- [ ] Mustard
- [ ] Oregano
- [ ] Parsley
- [ ] Spinach
- [ ] Thyme

FRUITS

- [ ] Frozen fruits
- [ ] Juneberries
- [ ] Rhubarb
- [ ] Strawberries

MEATS & FISH

- [ ] Beef
- [ ] Bison
- [ ] Chicken
- [ ] Emu
- [ ] Lamb
- [ ] Pork
- [ ] Rabbit
- [ ] Rainbow trout
- [ ] Sausages
- [ ] Smoked fish
- [ ] Turkey
- [ ] Venison
- [ ] Whitefish

EGGS & DAIRY

- [ ] Butter
- [ ] Buttermilk
- [ ] Cheese (cow, goat & sheep milk)
- [ ] Cottage cheese
- [ ] Cream
- [ ] Curds
- [ ] Dips
- [ ] Eggs
- [ ] Ice cream
- [ ] Milk
- [ ] Yogurt

OTHER

- [ ] Baked goods
- [ ] Beer
- [ ] Breads
- [ ] Cider
- [ ] Edible flowers
- [ ] Flours
- [ ] Ginseng
- [ ] Hickory nuts
- [ ] Honey
- [ ] Horseradish
- [ ] Jams
- [ ] Maple syrup
- [ ] Pesto
- [ ] Pickles & preserves
- [ ] Popcorn
- [ ] Prepared foods
- [ ] Salsa
- [ ] Sauces
- [ ] Spirits
- [ ] Sunflower oil
- [ ] Tortillas
- [ ] Vinegars
- [ ] Wild rice

Cheeses of the Month: Brie and Camembert

Profile: These are French-style soft-ripened cheeses shaped into flattened rounds and coated with a thin rind of white mold. Factory-made American Brie and Camembert, made from pasteurized milk, are generally less complex, more accessible than the originals. Brie producers drain the curds in metal hoops (which are containers with holes), then dry-salt and turn the cheese daily for a few days. Then they treat the rounds with bacteria and place them in cool, humid, ventilated rooms to allow the rind to bloom. Finally, the cheeses are ripened for several weeks. Camembert is made very similarly. Brie and Camembert consumers enjoy their soft-ripened favorites most often as table cheeses, but both types also serve splendidly as cooking ingredients.

Wisconsin brands include: Président Brie and Camembert, Woolwich Dairy USA's Goat Brie, Butler Farms' Camembert (sheep and cow milk), Caprine Supreme's La-Von (cow, goat and mixed milk Brie).

Taste: Mushrooms, earth, rich milk fat. Mild and sweet when young, growing increasingly funky and pungent with age.

Texture/Look: Creamy-soft and a bit shiny, with a firm, velvety, edible, snow-white rind. Young, commercial Brie and Camembert remain slightly firm but very tender at room temperature. Most classic styles "sag" as they warm and get oozy when very ripe and at room temperature; some even melt into a luscious, almost liquid, viscosity.

Go-withs: Chutney, cranberries, almonds, pecans, puff pastry, asparagus, spinach, sun-dried tomatoes, mushrooms, fresh herbs, cherries, plums, raspberries, strawberries, apples, dried fruit, ham, bacon, turkey.

What to drink: Sparkling wine, Kir royale, Sauvignon Blanc, Chardonnay, Pinot Noir, Sauternes, sherry, cherry kriek (a Belgian ale), apple or pear cider.

Classic dishes: Baked Brie, Brie en croûte, Camembert tarts.

Try this: Aspic-glazed Brie – Combine 2 cups dry white wine and 1 envelope unflavored gelatin in a saucepan; stir well and let stand 5 minutes. Place over medium flame and heat until mixture is clear. Place pan in a bowl filled with ice; stir until syrupy. Spoon some aspic over a round of Brie, then arrange edible decorations (herb sprigs, grape halves, carrot stars, capers, etc.) on top. Spoon on more aspic and let stand to set. Repeat with more aspic, if desired.

Photo ©Wisconsin Milk Marketing Board, Inc.

June 2014

Sunday	1
Monday	2
Tuesday	3
Wednesday	4
Thursday	5
	World Environment Day
Friday	6
Saturday	7
	REAP's Burgers & Brew

Risotto with Spinach, Green Onions, and Chives

Servings: 6–8

A bowl of creamy, steamy risotto is welcome most any time of year. This one features tender spring spinach, fresh herbs, and hand-grated Parmesan.

> 5–6 cups organic chicken stock
> 2 tablespoons butter
> ¼ cup finely chopped shallots
> 1 teaspoon dried thyme
> 1 ½ cups Arborio rice
> ⅔ cup dry white wine
> 6 cups fresh spinach leaves, cut into strips
> ¼ cup chopped green onions
> 1 cup fresh grated Parmesan, divided
> ¼ cup chopped fresh chives, divided
> 2 tablespoons butter, cut into small pieces and softened
> salt and pepper

Bring stock to a simmer in saucepan. Keep it hot while you heat 2 tablespoons butter in a large, heavy saucepan over a medium flame. When the butter begins to foam, add the shallots; reduce heat to medium-low and cook, stirring occasionally, until shallots are cooked, about 5 minutes. Add thyme and rice; stir 2–3 minutes to completely coat the rice. Raise heat to medium again and add the wine; stir and cook, stirring almost constantly, until nearly all the wine has evaporated, about 2 minutes.

Add two ladlefuls of the hot stock (enough to barely cover the rice); stir almost continuously until most of the stock is absorbed. Continue to add the stock, a ladleful at a time, and stir it almost constantly (or at least very frequently) until each ladleful is absorbed before adding the next. Stir in the spinach and green onions when nearly all the stock has been absorbed and the rice is suspended in a thick sauce. This will be a few minutes before the risotto is done. The rice is done when it's creamy-tender, which should take 30–35 minutes total. As it cooks, adjust heat if the rice is absorbing liquid too quickly.

When rice is done, fold in half the cheese, most of the chives, and the softened butter bits. Add salt and pepper to taste. Serve immediately, sprinkling each serving with more cheese and a few chives.

June 2014

Sunday	8
Monday	9
Tuesday	10
Wednesday	11
Thursday	12
Friday	13
Saturday	14

Flag Day

Sugar Snap Pea Salad with Radishes, Asiago, and Fresh Mint

Servings: 4–6

This is adapted from a recipe by New York Times *food writer Melissa Clark. One thing that's changed is how the peas and radishes are cut— they are angle-chopped, rather than sliced. Not only does this yield more crunch and juice than sliced vegetables, it looks more attractive. To make the angles, turn the vegetables after each cut to make small, spiky chunks.*

The cheese is also changed from the original recipe. In this version, Asiago lends a nutty contrast to the vegetables. We love the mint that Melissa used, but have found that dill makes a wonderful alternative, too.

> ¾ cup angle-chopped radishes
> 1 ½ cups angle-chopped sugar snap peas
> ¾ cup rough-chopped Asiago
> ¼ cup rough-chopped fresh mint leaves
> 1 teaspoon minced garlic
> sea salt
> 1 tablespoon fresh lemon juice
> 2–3 teaspoons balsamic vinegar
> 3 tablespoons extra virgin olive oil, or as desired
> freshly ground black pepper

Toss radishes, peas, cheese, and mint in a bowl. Using the flat of a knife, mash the garlic and pinch of salt on a cutting board to form a paste. Scrape garlic into a small bowl and add the lemon juice, vinegar, and a little more salt. Whisk in the olive oil. Toss with pea mixture. Add pepper to taste. Serve immediately or chill it first.

June 2014

Sunday	15
Father's Day	

Monday	16

Tuesday	17

Wednesday	18

Thursday	19
Juneteenth Day	

Friday	20

Saturday	21
First Day of Summer	

Strawberry Brie Blintzes

Servings: 8

Try this one for Sunday brunch; the crêpes can be made on Saturday and refrigerated overnight. Then all you have to do in the morning is fill and sauté them.

    6 eggs
    ½ teaspoon salt
    ¼ teaspoon nutmeg
    ⅓ cup plus ¼ cup flour
    1 cup water
    2 tablespoons vegetable oil, plus oil for cooking crêpes
    8 ounces chilled Brie, thinly sliced
    3 pints fresh strawberries, hulled and sliced (divided)
    sugar
    butter for cooking blintzes

To make batter, beat eggs with salt and nutmeg. Stir in flour. Stir in 2 tablespoons oil and 1 cup water until smooth (batter will be thin). If time permits, chill mixture 30–60 minutes.

To cook crêpes, heat a 7-inch nonstick skillet over a medium flame. Add a half-teaspoon oil and swirl to coat pan bottom. Discard hot oil. Stir crêpe batter; pour 3 tablespoons into the pan and immediately turn pan to coat the entire bottom. Cook until edges begin to curl—a minute or two—then loosen crêpe and flip it over. Cook on second side a few seconds; slide crêpe onto a plate. Repeat until batter is used up. If necessary, add a little water to keep batter thin.

To form blintzes, place 2–3 slices Brie and several sliced berries just below the center of a crêpe, keeping them level, not mounded. Fold bottom of crêpe over this, fold the two sides in, and then fold the whole thing up to create a small package with the seam side down. Continue until all the cheese is used up (but you should have some berries left). Cover and chill blintzes until you're ready to cook them. Combine remaining strawberries with a little sugar to taste and then purée.

To cook blintzes, place a parchment-lined baking sheet in a 200-degree oven. Heat a nonstick skillet over a medium flame. Add a little butter and swirl to coat pan. Cook blintzes in batches on both sides until blintzes are golden. As they finish, transfer them to the oven to keep warm. To serve: Arrange 2–3 blintzes on individual plates and drizzle with puréed berries..

June 2014

Sunday	**22**
Monday	**23**
Tuesday	**24**
Wednesday	**25**
Thursday	**26**
Friday	**27**

New Moon

Saturday	**28**

Ramadan Begins (Sunset)

Photo by Uriah Carpenter.

Chris Roelli, Cheesemaker and Co-owner

Roelli Cheese Company, Inc.
15982 Highway 11, Shullsburg
608-965-3779
Roellicheese.com

It's likely that the first real word Chris Roelli uttered as a toddler was "cheese," having entered the world as the probable fourth generation of Roelli cheesemakers. Later, when it would have been his turn to take over the family business, it seemed that he wouldn't have a chance to prove his mettle. Declining sales and a poor business climate had shuttered the factory in 1991, a fate befalling many other small cheesemakers around Wisconsin at that time. But Chris couldn't let go of his dream to be a cheesemaker. In 2005 he convinced his family to reopen their original cheese factory and re-equip it for small-scale production of artisan cheeses. The first cheese Chris made—a cave-aged, blue-veined Cheddar named Dunbarton Blue—was an instant winner, and it marked the start of a thriving artisan cheese business.

Please describe your cheeses. Our cheeses are small-batch, handmade, artisan products. We shelf-cure most of the cheese that we make. Most of our recipes are of the English style. We make two distinctly different Cheddar-blue recipes as our flagship offerings. We also do a few different varieties of Cheddar, a Cheshire, an Alpine-style, and a seasonal grass-based cheese.

What features make your cheeses stand out? We make small batches of cheese that allow us to use time-tested processes that require more labor than the large-scale producers are willing to invest. We also use the milk from one high-quality producer, allowing us the very best milk possible. All of our materials are natural—no artificial ingredients are used. Our milk source has pledged to not use artificial growth hormones. We also cure our cheese in a series of cellars to age them to perfection.

Which of the cheeses you make is your favorite? That's like asking which child is my favorite. They all present challenges that I like. Dunbarton was the first original recipe that I developed. It is probably the most special because it helped put our company in the area of the best cheese producers.

If you expanded your product line, what type of cheese that isn't well known in the U.S. would you make? Probably some of the small-scale mountain recipes from the Alps. Our Little Mountain cheese, which is similar to an Appenzeller, is an Alpine cheese.

What's your favorite thing about cheese? Cheese is a living thing. It is ever-changing. You have to pay attention to all the small details to make good cheese.

June 2014

Sunday	**29**
Monday	**30**
Tuesday	1
Wednesday	2
Thursday	3
Friday	4
Saturday	5

Gorgonzola hamburger. Photo courtesy of BelGioioso Cheese.

Kevin, Patrick, and Brian McCluskey are brothers, farmers, and partners of McCluskey Brothers at Shillelagh Glen Farms in Hillpoint. "We milk 63 cows and make our certified organic, 100 percent grass-fed cheese seasonally from May through November, when 'our girls' are on pasture. Hand crafted in small batches and made from our girls' milk only, we're known for our unpasteurized or 'raw milk cheeses,' but we also make some cheeses with pasteurized milk. 'One hundred percent grass-fed' means our livestock eat only two things their mom's milk and our pastures (or in winter, hay and haylage harvested from our fields). No grain, ever."

VEGETABLES

- [ ] Artichokes
- [ ] Beans
- [ ] Beets
- [ ] Bitter melon
- [ ] Bok choy
- [ ] Broccoli
- [ ] Cabbage
- [ ] Carrots
- [ ] Cauliflower
- [ ] Chinese cabbage
- [ ] Corn
- [ ] Cucumbers
- [ ] Dried chiles
- [ ] Fennel
- [ ] Garlic
- [ ] Kohlrabi
- [ ] Mushrooms (cultivated)
- [ ] Okra
- [ ] Onions
- [ ] Peas
- [ ] Potatoes
- [ ] Radishes
- [ ] Salad turnips
- [ ] Sprouts
- [ ] Squash blossoms & vines
- [ ] Summer squash
- [ ] Tomatoes
- [ ] Zucchini

HERBS & FRESH GREENS

- [ ] Chard
- [ ] Chives
- [ ] Cilantro
- [ ] Collards
- [ ] Dill
- [ ] Kale
- [ ] Kohlrabi
- [ ] Lettuces (leaf & head)
- [ ] Microgreens
- [ ] Mint
- [ ] Mustard greens
- [ ] Oregano
- [ ] Parsley
- [ ] Spinach
- [ ] Thyme

FRUITS

- [ ] Apricots
- [ ] Blueberries
- [ ] Cherries
- [ ] Currants
- [ ] Juneberries
- [ ] Mulberries
- [ ] Raspberries
- [ ] Strawberries

MEATS & FISH

- [ ] Beef
- [ ] Bison
- [ ] Chicken
- [ ] Emu
- [ ] Lamb
- [ ] Pork
- [ ] Rabbit
- [ ] Rainbow trout
- [ ] Sausages
- [ ] Smoked fish
- [ ] Turkey
- [ ] Venison
- [ ] Whitefish

EGGS & DAIRY

- [ ] Butter
- [ ] Buttermilk
- [ ] Cheese (cow, goat & sheep milk)
- [ ] Cottage cheese
- [ ] Cream
- [ ] Curds
- [ ] Dips
- [ ] Eggs
- [ ] Ice cream
- [ ] Milk
- [ ] Yogurt

OTHER

- [ ] Baked goods
- [ ] Beer
- [ ] Breads
- [ ] Cider
- [ ] Edible flowers
- [ ] Flours
- [ ] Ginseng
- [ ] Hickory nuts
- [ ] Honey
- [ ] Horseradish
- [ ] Jams
- [ ] Maple syrup
- [ ] Pesto
- [ ] Pickles & preserves
- [ ] Popcorn
- [ ] Prepared foods
- [ ] Salsa
- [ ] Sauces
- [ ] Spirits
- [ ] Sunflower oil
- [ ] Tortillas
- [ ] Vinegars
- [ ] Wild rice

Cheese of the Month: Colby

Profile: Invented in the town of Colby in the 1880s, Colby is a Wisconsin original. It's similar to young Cheddar, but it has a milder, sweeter taste because the curds are sprayed with cold water, which washes off the acidic whey. Colby is also moister and softer than its more famous cousin because of the washing, which helps prevent the curds from knitting  together, and because it isn't cheddared (see page 40). For old-style Colby, called longhorn Colby, the curds are packed into long cylinders and this yields Colby's characteristic nubby texture. However, due to legal changes in the definition of Colby that occurred during the 1970s, much Colby today is processed more quickly and simply than was traditionally done, and is considerably blander and smoother than classic Colby. The real McCoy is worth seeking out. Colby does not age well and should be eaten young. It can be sliced for sandwiches, cubed for snacks, and grated into casseroles. It also makes a classic Dairyland grilled cheese sandwich (don't stint on the pickles!).

Wisconsin brands include: Organic Valley's Certified Organic Colby, Decatur Dairy's ColbySwiss, Cady Creek Farms' Colby Longhorn, Dairymasters LLC's CoJack. Other highly regarded Colbys come from Gingerbread Jersey, Carr Valley Cheese, and Widmer's Cheese Cellars.

Taste: Sweet, mild, milky, simple.

Texture/Look: Slightly firm but giving, with an open, elastic texture perforated with tiny holes, creating a kind of marbled effect that is really pleasant on the tongue.

Go-withs: Fresh tomatoes, chiles, broccoli, apples, grapes, pickles, summer sausage, cold cuts, ham, smoked whitefish, burgers, macaroni, mustard, whole grain bread.

What to drink: Amber ale, dark lager, new Beaujolais, apple juice, milk.

Classic dishes: Grilled cheese sandwich, summer sausage and Colby sandwich, macaroni salad.

Try this: Locate some true longhorn Colby, slice it into thin rounds, and sandwich it with cotto salami between bread slices. Now cut around the cheese and meat, removing the extra bread and salami to make a perfectly round sandwich. Give the extra bread to the birds, and the salami scraps to the dog.

July 2014

Sunday 29

Monday 30

Tuesday 1

Wednesday 2

Thursday 3

Friday 4

Independence Day

Saturday 5

Papparadelle with Zucchini, Blue Cheese, and Tarragon
Servings: 4

Here's a yummy way to use some of that overflow of zucchini from the summer garden. For best flavor, pick the squash before they swell too large—about the width of two pinky fingers is just perfect (although bigger ones work, too). Braised with sautéed shallots and garlic and finished with fresh tarragon, this is a pasta dish that will have you planting another hill of zucchini next year.

Note: you can vary the cheese, herbs, and pasta shape. Think feta, dill, and penne, for example, or Parmesan, basil, and linguine. (Hmm, better make that two more hills next year.) The recipe directs you to cut fresh pasta into wide strips, but you can also substitute 4 to 6 ounces of cooked dried papparadelle (a wide, long flat pasta).

> 3 tablespoons olive oil
> ¼ cup thinly sliced shallots
> 3 medium garlic cloves, thinly sliced
> 1 ½ pounds small zucchini, diced
> salt and pepper
> 1 cup milk
> 1 package (12 ounces) fresh lasagna sheets, brought to room
>   temperature and each cut lengthwise into 8 strips
> 4–6 ounces coarsely crumbled blue cheese, divided
> 2 tablespoons chopped fresh tarragon, divided

Bring a large pot of salted water to boil. Meanwhile, heat olive oil gently in a large, heavy skillet over medium flame. Add shallots and garlic and cook, stirring occasionally, until softened, 3–4 minutes.

Add zucchini plus a generous seasoning of salt and pepper. Raise heat to medium and cook, stirring occasionally, until zucchini begins to soften, 5–10 minutes. Stir in the milk and continue to cook, stirring often, until zucchini is tender, about 10 minutes more.

While the zucchini cooks, add the pasta to the boiling water and cook until barely tender.

When the pasta is done, drain it and toss it with the zucchini, ¾ of the blue cheese, and ¾ of the tarragon. It will absorb any remaining liquid. Add more salt and pepper as needed. Divide into portions, sprinkle with remaining cheese and tarragon, and serve pronto.

July 2014

Sunday	6
Monday	7
Tuesday	8
Wednesday	9
Thursday	10
Friday	11
Saturday	12

Goat Cheese-Stuffed Squash Blossoms with Roasted Tomato Purée

Servings: 4–8

This is an elegant, special-occasion dish in which zucchini blossoms are filled with corn and chèvre, then fried and served with a roasted tomato purée. The recipe was contributed by Angie Carlson of Madison.

1 pint Sun Gold or other cherry tomatoes
¼ cup extra virgin olive oil
2 teaspoons balsamic vinegar, or more to taste
salt and pepper to taste
¾ cup flour
½ cup plus 2 tablespoons beer
1 large egg
6 ounces fresh chèvre, at room temperature
¼ cup pine nuts or walnuts
½ cup fresh corn kernels, divided
2 tablespoons chopped basil, divided
12–16 large zucchini or squash blossoms, stems and stamens removed
olive oil for frying

Heat oven to 250 degrees. Line a baking sheet with parchment paper. Halve the tomatoes and arrange cut-side up in a single layer on the sheet. Roast for about 2 hours, checking occasionally to make sure they aren't browning. Transfer to a blender; add olive oil and vinegar. Blend to form a thick sauce, adding a little water if necessary. Season with salt, pepper, and a little more vinegar to taste.

While tomatoes are roasting, combine flour and beer, then whisk in the egg. Set batter aside as you prepare the filling and stuff the blossoms: Place chèvre and pine nuts in a bowl. Reserve a few corn kernels and basil leaves for garnish, then add the rest to the bowl; mix ingredients well. Gently spread apart each blossom and stuff it with a heaping teaspoon of filling, pushing the filling towards the base and twisting the upper part of the flower to seal in the filling.

Add olive oil to a depth of ½ inch in a large skillet. Heat over a medium-high flame until shimmering. Dip blossoms into batter and fry without overcrowding them, turning periodically to ensure even browning. When golden-brown, drain them on paper towels. Arrange 2 to 4 blossoms atop some sauce on serving plates. Garnish with reserved corn and basil.

July 2014

Sunday	13
Monday	14
Tuesday	15
Wednesday	16
Thursday	17
Friday	18
Saturday	19

Blueberry Ice Box Torte

Servings: 12–16

*Tortes are a big thing in Wisconsin. We like the classy European ones like
Linzer torte and Napoleon torte well enough, but what we really go for
are the homier, cream cheese-y ones, the kind that start with a graham
cracker or cookie crust, get layered with fruit or pudding, and are topped
with a mass of whipped cream. Here's one of those beauties.*

> 3 cups graham cracker crumbs
> 6 tablespoons butter, melted
> 2 pounds blueberries (if frozen, thaw them first)
> ⅔ cup sugar
> grated zest of 1 orange
> ½ cup plus 3 tablespoons water, divided
> 4 tablespoons cornstarch
> 2 packages (each 8 ounces) cream cheese, softened
> 1 cup plus 2 tablespoons powdered sugar, divided
> 3 tablespoons milk
> 1 ½ teaspoons vanilla extract, divided
> 2 cups heavy cream

Combine graham cracker crumbs and butter. Press mixture into bottom
of a 9-by-13-inch aluminum pan. Freeze crust at least 2 hours.

Combine berries, sugar, orange zest, and ½ cup of the water in a
saucepan (if you're using thawed berries, skip the water but include
any berry juice that's present). Bring to a simmer and cook until
sugar dissolves, 1-2 minutes. Combine cornstarch with remaining 3
tablespoons water and stir most of it into the blueberries. The mixture
will quickly thicken. If it's not very thick, add more cornstarch mixture.
Give the mixture a few more stirs, cool it to room temperature, then chill
it thoroughly.

Beat cream cheese and 1 cup powdered sugar with electric beaters
until well-combined. Beat in milk and 1 teaspoon vanilla. Clean and dry
beaters. Using another bowl, whip cream at medium speed until thick
and fluffy. Gradually blend in remaining powdered sugar and vanilla,
raise the speed, and beat until stiff peaks form.

Spread cream cheese filling evenly over crust. Spread fruit evenly over
cream cheese. Spread whipped cream over fruit, or arrange it in dollops
to suggest serving sizes. Chill torte until serving time.

July 2014

Sunday	**20**
Monday	**21**
Tuesday	**22**
Wednesday	**23**
Thursday	**24**
Friday	**25**
Saturday	**26**
	New Moon

Sid Cook, Owner and Master Cheesemaker

Carr Valley Cheese Company
S3797 County Highway G, La Valle
608-986-2781
carrvalleycheese.com and artisancheesesource.com

Fourth-generation cheesemaker Sid Cook had it easy. When he was young, he only had to open a door off the family kitchen and he was in the make room of his family's cheese plant, the room where science and art come together and transform milk to cheese. So it isn't hard to see why Sid became enamored with cheese production at an early age. He was all of twelve when he began working with cheese and obtained his cheesemaker's license at sixteen. In the mid-1970s Sid and his brother took over the family business, but as it became harder to make a living producing "standard" cheeses, Sid ventured into something new. He purchased the Carr Valley Cheese Company, a small factory in La Valle that had been in existence since 1902, and started to craft artisan cheeses using milk from cows, sheep, or goats, as well as mixed milks. Carr Valley now makes more than 80 different cheeses. Sid masterminds an operation that consists of three plants and eight retail stores. Along the way he became a Wisconsin Master Cheesemaker with certificates in Cheddar and Fontina. His one-of-a-kind proprietary cheeses known as American Originals are particularly award-worthy and have helped make Sid the most decorated cheesemaker in North America.

Where do you get your inspiration to craft a new cheese? I imagine what can be.

For each of the four seasons of the year, what cheese do you like best? Spring: Irish Valley Cheddar, made just before the cows go on grass. Summer: Wild Flower Cheddar, when the cows are on grass and wildflowers. Fall: Autumn Harvest Cheddar, smoked with apple wood. Winter: Winter Solstice Cheddar, a white Cheddar.

Could you blind-taste your cheese in a lineup of similar cheeses? Hopefully. I would for sure know which ones that I wished that I had made.

What is it about cheese that makes it an ideal ingredient to cook with? A small amount can make a big difference in flavor and really make the dish special.

Why is Wisconsin the only state to offer a Master Cheesemaker program? I guess it's because we're a couple of light years ahead.

July 2014

	Sunday	27
	Ramadan Ends	
	Monday	28
	Tuesday	29
	Wednesday	30
	Thursday	31
	Friday	1
	Saturday	2

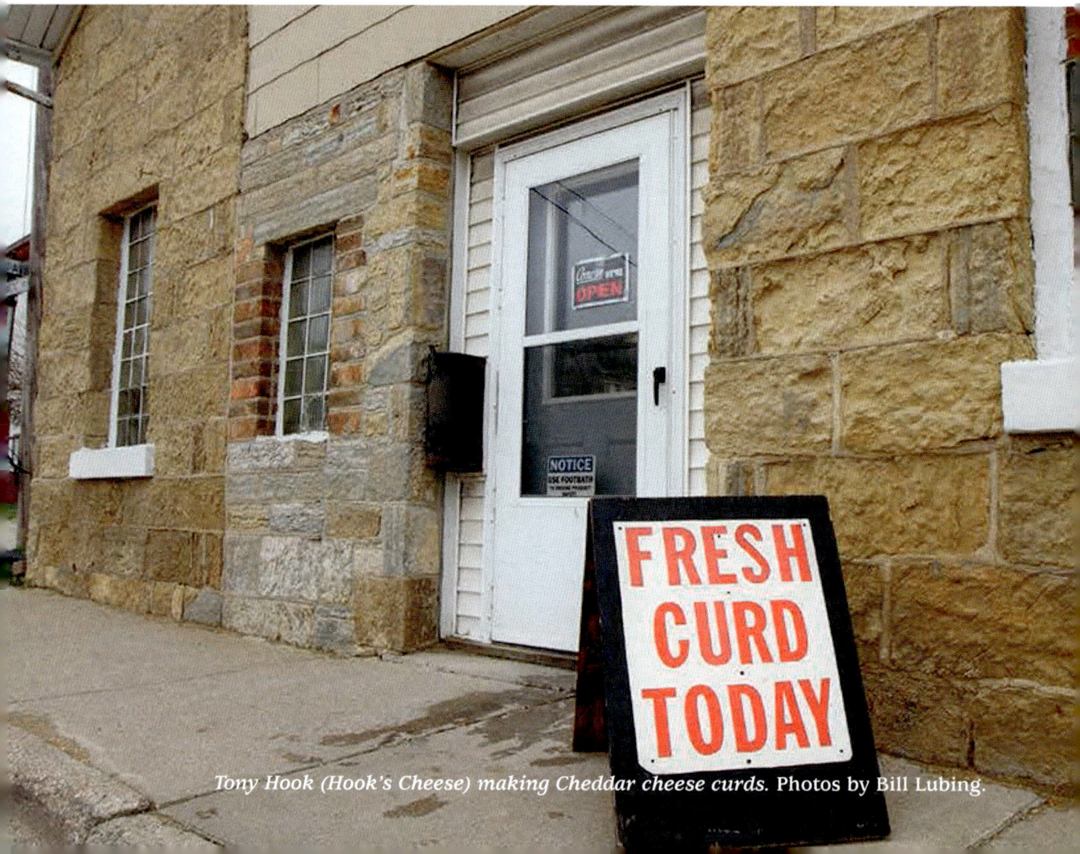

Tony Hook (Hook's Cheese) making Cheddar cheese curds. Photos by Bill Lubing.

If the best hot-weather cooking is no cooking at all, then the best recipe is, of course, no recipe. Ripe fruit and cheese layered on fresh bread is as easy and tempting as summer food gets. Try melon, mint, and ricotta on brioche, or raspberries, walnuts, and cream cheese on a croissant. Consider Muenster and thinly-sliced pears on cracked wheat bread, or rye bread with feta butter and heirloom tomatoes.

VEGETABLES

- Artichokes
- Beans
- Beets
- Bitter melon
- Carrots
- Cauliflower
- Celery
- Corn
- Cucumbers
- Eggplant
- Fennel
- Garlic
- Kohlrabi
- Leeks
- Mushrooms (cultivated)
- Okra
- Onions
- Peppers
- Potatoes
- Summer squash
- Tomatillos
- Tomatoes
- Zucchini

HERBS & FRESH GREENS

- Chard
- Cilantro
- Collards
- Dill
- Kale
- Kohlrabi
- Lettuces (leaf & head)
- Mustard greens
- Oregano
- Parsley
- Rosemary
- Sage
- Spinach
- Tarragon
- Thyme

FRUITS

- Apples
- Apricots
- Blackberries
- Blueberries
- Cherries
- Currants
- Elderberries
- Melons
- Mulberries
- Peaches
- Plums
- Raspberries
- Strawberries

MEATS & FISH

- Beef
- Bison
- Chicken
- Emu
- Lamb
- Pork
- Rabbit
- Rainbow trout
- Sausages
- Smoked fish
- Turkey
- Venison
- Whitefish

EGGS & DAIRY

- Butter
- Buttermilk
- Cheese (cow, goat & sheep milk)
- Cottage cheese
- Cream
- Curds
- Dips
- Eggs
- Ice cream
- Milk
- Yogurt

OTHER

- Baked goods
- Beer
- Breads
- Cider
- Edible flowers
- Flours
- Ginseng
- Hickory nuts
- Honey
- Horseradish
- Jams
- Maple syrup
- Pesto
- Pickles & preserves
- Popcorn
- Prepared foods
- Salsa
- Sauces
- Spirits
- Sunflower oil
- Tortillas
- Vinegars
- Wild rice

Cheese of the Month: Feta

Profile: Developed first in Greece from sheep and goat milk, feta is one of the world's oldest cheeses. In the United States, it's usually made from pasteurized cow milk, although artisan cheesemakers use other milks, too. Feta is a fresh cheese that is brined after it's been formed and thus can last months longer than other fresh varieties. (To reduce the saltiness, rinse or soak it in water before serving.) It can also be aged for up to a year. Although it's not a slicing or melting cheese, feta is marvelously versatile in salads, egg dishes, and pastas; atop burgers, grains, and pizzas; and with countless vegetables. Fresh herbs and spices love feta; indeed, it often comes flavored with the likes of dill, chives, and pepper.

Wisconsin-made styles include: Greek, French, Israeli, Italian, and Bulgarian styles, goat milk, cow milk, sheep milk, or a mixture of milks. Top-notch Wisconsin artisan and specialty brands include Butler Farms, Dreamfarm, Mt. Sterling Co-op Creamery, Capri, Trega Foods, Klondike Cheese Company, and Hillbilly Hollow.

Taste: Salt, tang, tartness. The goat and sheep varieties are tangier and pleasantly barn-y.

Texture: Soft, moist, and creamy-crumbly, to tender-firm, drier, and very crumbly.

Go-withs: Spinach, garlic, tomatoes, edamame, cucumbers, green beans, eggplant, zucchini, sweet red peppers, garlic, melon, oranges, lemon, lamb, chicken, tabouli, fresh herbs, olive oil, honey, bread, pasta.

What to drink: Pinot Noir, Zinfandel, Sauvignon Blanc, Pinot Blanc, Chenin Blanc, wheat beer, pilsner, lager, fruit beer, vodka, green tea.

Classic dishes: Greek salad, spanakopita, pastitsio.

Try this: Arrange the following in small piles on a platter: grilled eggplant and zucchini, roasted red peppers, hard-cooked eggs, boiled Yukon gold potatoes, cherry tomatoes, oil-cured black olives, and grilled lamb sausage. Scatter crumbled feta around the piles. Drizzle everything with a dressing that contains chopped cilantro and parsley, minced garlic, lemon juice, and olive oil.

August 2014

Sunday	27
Monday	28
Tuesday	29
Wednesday	30
Thursday	31
Friday	1
Saturday	2

Bubbly Cheese on the Grill

Servings: 12 or more

This appetizer won a prize at the 2002 Food for Thought Festival in Madison. For years this annual contest, sponsored by REAP Food Group (see page 200), challenged cooks to "go local" with their entries and helped spread the word that cooking sustainably can be easy, fun, and delicious. Apparently the word spread at least as far as Orlando, Florida, because that's where the contributor of this recipe, Radalle Knappenberger, is from!

8 pita pockets, cut into bite-size wedges (or substitute store-bought corn chips)
2 tablespoons olive oil
2 medium onions, chopped
3 cloves garlic, minced or mashed
½ cup chopped sun-dried tomatoes
½ cup chopped kalamata or other black olives, chopped
2 fresh jalapeños, seeded and minced
2 pounds Wisconsin Colby cheese
4 apples, each cut into 8 wedges (optional)

If desired, toast pita wedges on a lightly greased cookie sheet in a 350-degree oven for 10 minutes, or until well toasted.

Heat olive oil over a medium-high flame in a medium skillet. Add onions and garlic and cook until tender, 8–10 minutes. Remove from heat and stir in tomatoes, olives, and jalapeños. Cut cheese into medium chunks and place in a 9- or 10-inch flameproof baking dish that's at least 1 ½ inches deep. Spread onion-tomato mixture over cheese.

Heat an outdoor grill. Place baking dish on grill about 6 inches above heat. Grill just until the cheese melts, then move to a cooler place on the grill. Scoop up the cheese with the pita wedges or corn chips. Serve with apple wedges, if desired.

August 2014

Sunday	3
Monday	4
Tuesday	5
Wednesday	6
Thursday	7
Friday	8
Saturday	9

Watermelon Feta Salad

Servings: 5–6

Crunchy, juicy, sweet, salty, and zingy: this salad has it all. And who knew watermelon and feta (and basil!) would take to each other so well. The recipe is adapted from one on the National Watermelon Promotion Board website.

Hint: Use the inner white "stems" of the romaine as well as the leaves; they're part of what add the juice and crunch.

6–7 cups chopped romaine lettuce, washed and spun dry
  in a salad spinner
3 cups seeded, cubed watermelon
½ cup very thinly sliced red onion
½ cup crumbled feta
¼ cup puréed watermelon
2 tablespoons finely sliced fresh basil leaves
2 tablespoons balsamic vinegar
salt and freshly ground black pepper to taste
2 tablespoons extra virgin olive oil
flaked sea salt

Place salad greens in a light cotton towel and roll it up (but not tightly). Place watermelon cubes, red onions, and feta in three containers. Chill all four ingredients until just before serving time.

For dressing, mix puréed watermelon, basil, vinegar, salt, and pepper in a small bowl. Whisk in olive oil a little at a time. Taste for seasoning and add more salt and pepper if needed.

Just before serving, place the lettuce in a large, bright-colored salad bowl. Add the other chilled ingredients (drain the watermelon cubes first, if any liquid has accumulated). Toss with just enough dressing to lightly coat everything. Portion onto serving plates, sprinkle with a tiny amount of flaked sea salt, and serve without delay.

August 2014

Sunday	10
Monday	11
Tuesday	12
Wednesday	13
Thursday	14
Friday	15
Saturday	16

Savory Tomato, Bread and Cheese Pudding

Servings: 8–12

The contributor of this recipe, Amihan Huesman, served this dish for the first time on the Sunday morning after September 11, 2001, when she wanted to gather friends close. It's at its comforting best, of course, when tomatoes are in season.

6 eggs, beaten
2 ½ cups milk or 3 cups buttermilk
2–3 tablespoons snipped fresh chives
salt and pepper to taste
12 slices white or whole wheat sandwich bread (about ¾
  pound), crusts sliced off
2 tablespoons butter, plus more for the baking dish
1–1 ½ cups diced onion
2 tablespoons pesto (optional)
6 medium tomatoes (2–3 inches in diameter), trimmed, seeded,
  and sliced
½ – ¾ pound shredded cheese (Swiss, Cheddar or any good
  melting variety you like), divided

Heat oven to 350 degrees. Mix the eggs, milk, and chives in a large, wide bowl. Season the mixture with salt and pepper. Add the bread and let it soak, pressing the slices down to make them soak up the egg mixture. Heat 2 tablespoons of the butter in a skillet over a medium flame, add onions, and cook until golden and wilted. Remove onions from heat and stir in the pesto, if using.

Butter a 9-by-13-inch baking pan. Layer the ingredients in the pan in this order:

½ the bread
½ the tomatoes
all of the onion-pesto mixture
¾ of the cheese
½ the bread
½ the tomatoes

Cover pan with foil and bake 20 minutes. Uncover and bake until bread pudding is puffed and beginning to brown, 20–25 minutes longer. Sprinkle the last of the cheese on top and return to the oven until the cheese is melted. Let stand 10–15 minutes before serving.

August 2014

	Sunday	17
	Monday	18
	Tuesday	19
	Wednesday	20
	Thursday	21
	Friday	22
	Saturday	23

Brenda Jensen, Owner and Cheesemaker

Hidden Springs Creamery
S1597 Hanson Road, Westby
608-634-2521
hiddenspringscreamery.com

Brenda Jensen once was a reluctant sheep farmer, coaxed into the role by her husband, a wannabe sheep dairyman from the city, after he deposited 50 sheep on her (farm) doorstep. But once Brenda discovered the quality of her sheep herd's milk, she imagined making some cheese, and it soon became obvious how this would play out. Brenda took a cheesemaking class and became hooked—thank goodness! Her success can be measured by the many awards she has received for her various cheeses.

Tell us about your cheeses. The first cheese we made was a creamy, soft, fresh, sheep's milk cheese. We followed the plain version with an assortment of flavored ones that continues to evolve. Others in our family of distinctive cheeses include raw milk, washed-rind varieties; semi-soft blue cheeses; and feta.

What features make your cheeses stand out? We are a farmstead sheep milk operation. All of the milk comes from our own herd. We do not use chemicals, pesticides, or antibiotics, and are growth hormone-free.

What are unique ways to serve, cook with, or otherwise prepare your cheeses? Our cheeses are great with crackers and bread. You can also grate and melt these cheeses. Our soft cheese can be used in place of chèvre or ricotta in most recipes. You can use this cheese for pizza, lasagna, dips, in omelets, or as a pasta sauce.

Which of the cheeses you make is your favorite? I am a mood cheese eater. I love all of our cheeses, and eat them on at least a weekly basis!

What's your favorite thing about cheese? My favorite thing about our cheese is we can take great care of our animals and they will take great care of us by producing high quality milk, which we turn into great-tasting cheese.

What hooked you most when you were learning your craft? I fell in love with the smells of the milk warming and the magic of cheesemaking.

August 2014

Sunday	**24**
Monday	**25**
New Moon	
Tuesday	**26**
Wednesday	**27**
Thursday	**28**
Friday	**29**
Saturday/Sunday	**30** / **31**

Felix Thalhammer (Capri Cheese) offering up a big cheesy smile. Photo by Bill Lubing.

"We are seasonal producers, so we do not craft cheese in the winter," says Diana Kalscheur Murphy of Dreamfarm in Cross Plains. Diana specializes in fresh-style farmstead cheese made from pasture-based goat milk. "Because we are seasonal, our milk is constantly changing through the goats' stages of lactation and the change in weather and environment." Is there one timeframe that stands out? "I don't have a cheese I like best as we go through these changes; I enjoy the subtle changes as they occur. Although early fall produces a very creamy fresh cheese."

VEGETABLES

- Beans
- Beets
- Broccoli
- Brussels sprouts
- Cabbage
- Carrots
- Cauliflower
- Celeriac
- Celery
- Chinese cabbage
- Corn
- Cucumbers
- Dried beans
- Dried chiles
- Eggplant
- Garlic
- Green tomatoes
- Kohlrabi
- Leeks
- Mushrooms (cultivated)
- Onions
- Parsnips
- Peppers
- Potatoes
- Pumpkins
- Radishes
- Rutabaga
- Summer squash
- Sweet potatoes
- Sweet red peppers
- Tomatillos
- Tomatoes
- Turnips
- Winter squash
- Zucchini

HERBS & FRESH GREENS

- Chard
- Chives
- Cilantro
- Collards
- Kale
- Kohlrabi
- Lettuces (leaf & head)
- Microgreens
- Mustard greens
- Oregano
- Parsley
- Rosemary
- Sage
- Spinach
- Tarragon
- Thyme

FRUITS

- Apples
- Apricots
- Concord grapes
- Cranberries
- Melons
- Peaches
- Pears
- Plums
- Raspberries

MEATS & FISH

- Beef
- Bison
- Chicken
- Emu
- Lamb
- Pork
- Rabbit
- Rainbow trout
- Sausages
- Smoked fish
- Turkey
- Venison
- Whitefish

EGGS & DAIRY

- Butter
- Buttermilk
- Cheese (cow, goat & sheep milk)
- Cottage cheese
- Cream
- Curds
- Dips
- Eggs
- Ice cream
- Milk
- Yogurt

OTHER

- Baked goods
- Beer
- Breads
- Cider
- Edible flowers
- Flours
- Ginseng
- Hickory nuts
- Honey
- Horseradish
- Jams
- Maple syrup
- Pesto
- Pickles & preserves
- Popcorn
- Prepared foods
- Salsa
- Sauces
- Spirits
- Sunflower oil
- Tortillas
- Vinegars
- Wild rice

Cheeses of the Month: Latino Styles

Profile: Every time a cheesemaking immigrant group has settled in Wisconsin, the dairy scene has benefited. We usually associate Wisconsin cheese with European traditions, but Hispanic varieties also play an important role; Mexican cuisine in particular has added countless tasty cheese varieties to the mix. They generally come in three styles: fresh, melting, and hard. When heated, the crumbly, white, fresh varieties soften but keep their shape without melting, making them the cheese of choice to garnish hot foods like refried beans and chilaquiles, or to fill stuffed dishes like chiles rellenos and enchiladas. Queso Blanco, Panela, Queso Fresco, and other fresh cheeses are also excellent on salads and with fruit. Asadero and Queso Quesadilla are two melting cheeses, crafted to succumb smoothly and "grease-lessly" to heat; they are favorites in tacos, quesadillas, and nachos. Hard varieties include Cotija and Enchilado; these are aged, grating cheeses used much like Parmesan.

Wisconsin brands include: Chula Vista Cheese Company, Specialty Cheese Company, Decatur Dairy, Valley View Cheese, Wisconsin Cheese Group, Cesar's Cheese.

Taste: From rich and mild or lightly tangy, to strong, sharp and salty.

Texture/Look: Fresh styles are firm, white, and crumbly; melting types can be smoother; and hard varieties are extra-firm.

Go-withs: Melons, peaches, tropical fruits, chiles, tomatoes, avocados, poultry, beef, pork, dry beans, cornmeal, salsa, cilantro, oregano, cumin, tortillas.

What to drink: With fresh and melting varieties try white wine, sangria, lager, bock, margaritas, fruit juice, and milk. With hard aged cheeses, go for Zinfandel or dark lager.

Classic dishes: Enchiladas, chiles rellenos, tamale pie, huevos rancheros, nachos, burritos.

Try this: Sweet Corn Cheese Wedges – Combine 3 heaping cups sweet corn kernels, 1–3 minced serranos, 8–10 ounces shredded Asadero, 5 beaten eggs, and some chopped cilantro. Spread in an oiled ceramic dish and bake in preheated 350-degree oven until set, 35–45 minutes. Let stand 10 minutes before serving.

September 2014

Saturday/Sunday 30 31

Monday 1

Labor Day

Tuesday 2

Time to order your 2015 Wisconsin Local Foods Journal

Wednesday 3

Thursday 4

Friday 5

Saturday 6

Pine Nut Couscous with Arugula, Sweet Peppers, and Feta
Servings: 4–6

*A first-place winner in the 2000 Food for Thought Recipe Contest
sponsored by REAP Food Group, this imaginative salad comes from
Friedie Carey of Middleton, Wisconsin.*

> 2 ½ tablespoons red wine vinegar
> 1 tablespoon Dijon-style mustard
> salt and pepper to taste
> ¼ cup olive oil
> 1 box (5.6 ounces) toasted pine nut couscous
> ⅔ cup crumbled feta cheese (plain or studded with peppercorns),
>   divided
> ½ cup each red and yellow sweet pepper, thinly sliced
> 1 ½ cups sliced arugula leaves
> 2 chopped green onions
> 1 large tomato, chopped
> 3 tablespoons capers
> 2 medium garlic cloves, minced
> salt and pepper to taste
> 2 tablespoons pine nuts, toasted a few minutes at 350 degrees

Combine vinegar, mustard, and some salt and pepper in a medium
bowl; whisk in olive oil. Prepare couscous as directed on package. Toss
with dressing while couscous is still warm; cool to room temperature
or until slightly warm.

Set aside 2 tablespoons of the crumbled feta. Toss couscous with the
larger amount of feta, plus the sweet peppers, arugula, green onions,
tomato, capers, garlic, and salt and pepper to taste. Place in serving
bowl, and sprinkle with toasted pine nuts. Serve immediately or chill
until ready to serve.

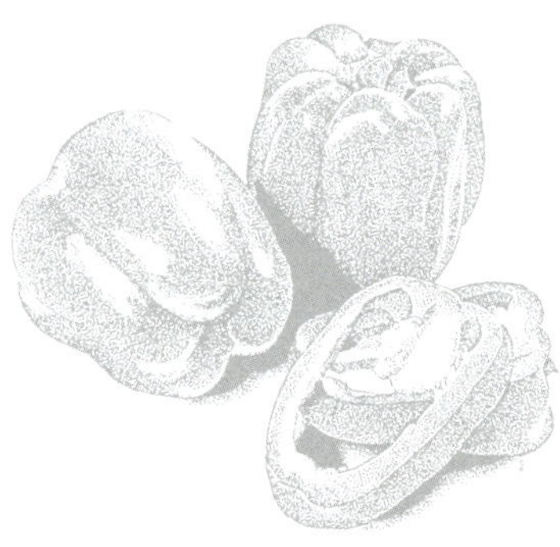

September 2014

Sunday	7
Monday	8
Tuesday	9
Wednesday	10
Thursday	11
Friday	12
Saturday	13

Bacon, Blue Cheese, and Pear Bites

Servings: 10–12

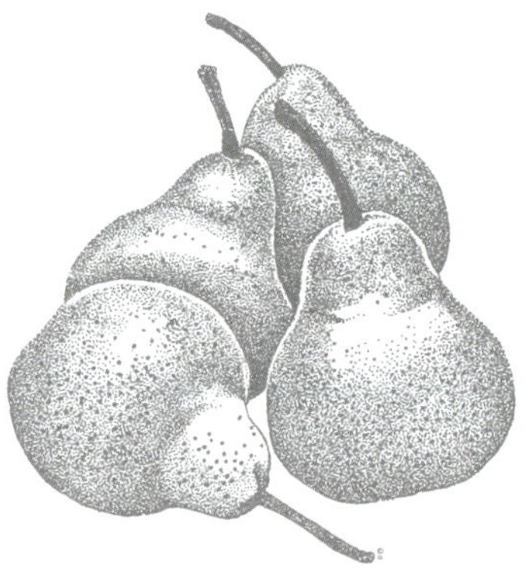

Blue cheese and pears are a classic sweet-and-salty combination that never goes out of style. Add a little smoky bacon to the mix, bake it on sliced French bread, and classic becomes cutting edge.

Note: Peeled, sliced pears can quickly turn brown, so if you're not going to bake these immediately, toss the pears with some lemon juice before assembling the little stacks. Cover with plastic wrap until you're ready to bake them.

    1 baguette or slender loaf of French bread
    3–4 ripe pears
    softened butter
    ⅔ cup pear butter*
    4 ounces crumbled blue cheese
    ¾ cup cooked, crumbled bacon

Heat oven to 400 degrees. Use a serrated knife to cut the bread on the diagonal into slices that are about ½ inch thick. Arrange the sliced bread in a single layer on a baking sheet.

Peel the pears, cut them in half lengthwise, pull out the stems, and scoop out the cores. Thinly slice the halves lengthwise. Spread one side of each slice of bread with a little softened butter and then some pear butter. Arrange 3–4 pear slices across each one and sprinkle it with blue cheese and bacon crumbles.

Bake until cheese is melted and everything is hot, 8–10 minutes. Serve immediately.

*Pear butter is a thick spread made from cooked pears. If it's not available, substitute apple butter.

September 2014

Sunday	14
Monday	15
Tuesday	16
Wednesday	17
Thursday	18
Friday	19
Saturday	20

REAP's Food for Thought Festival

Chiles Rellenos

Serves: 3

Poblanos, with their thick walls and lightly spicy kick, are the peppers of choice when making chiles rellenos. To keep things authentic, use a Mexican melting cheese such as Queso Quesadilla or Chihuahua for their mildness, or go for Asadero, which is a bit stronger. If none of these are available (or if authenticity isn't a goal), you'll also do just fine with Monterey Jack, mozzarella, or a young Brick.

> 6 large fresh poblanos
> 4-5 ounces melting cheese, sliced
> ½ cup flour
> 1 teaspoon salt
> vegetable oil
> 3 eggs, separated
> 2 cups puréed spicy tomato-chile sauce (homemade or
>   purchased), heated
> fresh cilantro sprigs

Broil or roast poblanos at 500 degrees, turning them occasionally, until skins are blackened and blistered all over. Place in closed paper bag for 10 minutes to help loosen the skins. Gently peel off skins. Make a slit along one side of each chile. Carefully remove seeds but do not remove stems. Place 1 to 1 ½ ounces cheese inside each chile and then close it up. Mix flour and salt and reserve 2 teaspoons of it. Gently roll stuffed peppers in the remaining flour mixture.

Heat oil to about a ½-inch depth in a cast iron pan. Beat 3 egg whites in a medium bowl until stiff. Using a second larger, bowl, mix the reserved 2 teaspoons of the flour mixture with the 3 egg yolks. Gently fold in the egg whites to make a batter.

Heat 3 serving plates in a warm oven while you fry the chiles.

Check heat of oil—it should be hot enough to make a little bit of the batter sizzle when it hits the oil. Dip each chile into the egg batter, turning to coat it well. Fry chiles in the hot oil, turning them once, until golden brown all over. Don't crowd the pan; do this in batches if necessary.

Portion the tomato sauce onto the warm serving plates. When chiles are done, drain them briefly on paper towels and portion up 2 per plate. Garnish with cilantro and serve immediately.

September 2014

Sunday	**21**
Monday	**22**
Tuesday	**23**

First Day of Fall

Wednesday	**24**

Rosh Hashanah Begins (Sunset)
New Moon

Thursday	**25**
Friday	**26**

Rosh Hashanah Ends (Sunset)

Saturday	**27**

Ken Heiman, General Manager, Partner and Master Cheesemaker

Nasonville Dairy
10898 U.S. Highway 10 West, Marshfield
715-676-2177
nasonvilledairy.com

Nasonville Dairy was founded in 1885, and since 1965 the Heiman family has been the face of it. That is when cheesemaker Arnie Heiman and his wife Rena Mae began managing the plant. Twenty years later they owned it. Like their father, sons Ken, Kim, and Kevin became licensed cheesemakers—Ken at the age of 16—and all three brothers, as well as some of their children, are actively involved with the business today. The plant has seen tremendous growth in the time that the Heimans have taken ownership in 1985. At that time, Nasonville was taking in 74,000 pounds of milk per day and now the dairy processes 1.4 million pounds of milk per day.

Ken went on to become a Master Cheesemaker with certificates in Feta and Monterey Jack. He even went to Greece to learn more about making feta. Ken is now working towards Master Cheesemaker certifications in Cheddar and Asiago cheese varieties.

Please describe your cheeses. We make some 30-plus different kinds of cheeses, from the elementary (Cheddar, Colby, Pizza Cheese, Feta, etc.) to a wide variety including Italian (Asiago, Provolone, Parmesan), flavored Monterey Jack styles (Blue Marble Jack, Buffalo Wing, Horseradish, Jalapeño Pepper, Chipotle, etc.), Mexican/Hispanic (Queso Blanco and Cotija), and more.

What features make your cheeses stand out? Nasonville is fortunate in that we don't "make cheese" to make cheese, we make it for our specific customers, so we are able to accommodate their needs/requests more easily. We offer kosher cheese, recombinant bovine growth hormone-free cheese, cuts from 40-pound blocks to five-pound and ten-pound loaves, to crumbled, to diced… and that's not all. We are also bringing something new to the marketplace. Nasonville is the only plant in Wisconsin to offer a cheese with five times the levels of naturally occurring omega-3 fatty acids without adding fish oil.

Which of the cheeses you make is your favorite and tell us why? Garlic & Herb Jack. It has a great aroma, a nice flavor profile throughout, and is very versatile to use.

Where do you get your inspiration to craft a new cheese? Some of the inspiration comes from personal world travels, while other ideas actually come right from a customer or employee saying, "I wonder what this would taste like." Nasonville is not afraid to try new things and our customers know that they, too, can come in with a special request and we'll see if we can help them out.

September 2014

	Sunday	28
	Monday	29
	Tuesday	30
	Wednesday	1
	Thursday	2
	Friday	3
	Saturday	4

The possibilities for seasonal pairings with cheese are nearly endless. Here's some recommendations for fall, from people who know best:

TONY HOOK, HOOK'S CHEESE:
"A sharp **Cheddar**, like Hook's ten-year Cheddar, with apples or pears."

VICKI THINGVOLD, MEISTER CHEESE:
"Three Alarm **Colby Jack** has a nice heat for tailgating and parties."

PAULA HOMAN, RED BARN FAMILY FARMS:
"Heritage Weis Reserve, [an aged white **Cheddar**], is creamy, flavorful comfort food that's even better when used for mac and cheese or sprinkled on chili, tacos, or homemade French fries."

FRANCIS WALL, BELGIOIOSO:
"We like to grill out for the Packer games and nothing tastes better on a steak or burger than our CreamyGorg, [a **Gorgonzola**]."

VEGETABLES
- [ ] Beets
- [ ] Broccoli
- [ ] Brussels sprouts
- [ ] Burdock root
- [ ] Cabbage
- [ ] Carrots
- [ ] Cauliflower
- [ ] Celeriac
- [ ] Chinese cabbage
- [ ] Dried beans
- [ ] Dried chiles
- [ ] Garlic
- [ ] Green tomatoes
- [ ] Kohlrabi
- [ ] Leeks
- [ ] Mushrooms (cultivated)
- [ ] Napa cabbage
- [ ] Onions
- [ ] Parsnips
- [ ] Potatoes
- [ ] Pumpkins
- [ ] Rutabaga
- [ ] Sweet potatoes
- [ ] Turnips
- [ ] Winter radishes
- [ ] Winter squash

HERBS & FRESH GREENS
- [ ] Kale
- [ ] Kohlrabi
- [ ] Lettuce
- [ ] Microgreens
- [ ] Oregano
- [ ] Parsley
- [ ] Rosemary
- [ ] Sage
- [ ] Spinach
- [ ] Thyme

FRUITS
- [ ] Apples
- [ ] Cranberries
- [ ] Frozen fruits
- [ ] Pears
- [ ] Raspberries

MEATS & FISH
- [ ] Beef
- [ ] Bison
- [ ] Chicken
- [ ] Emu
- [ ] Lamb
- [ ] Pork
- [ ] Rabbit
- [ ] Rainbow trout
- [ ] Sausages
- [ ] Smoked fish
- [ ] Turkey
- [ ] Venison
- [ ] Whitefish

EGGS & DAIRY
- [ ] Butter
- [ ] Buttermilk
- [ ] Cheese (cow, goat & sheep milk)
- [ ] Cottage cheese
- [ ] Cream
- [ ] Curds
- [ ] Dips
- [ ] Eggs
- [ ] Ice cream
- [ ] Milk
- [ ] Yogurt

OTHER
- [ ] Baked goods
- [ ] Beer
- [ ] Breads
- [ ] Cider
- [ ] Flours
- [ ] Hickory nuts
- [ ] Honey
- [ ] Horseradish
- [ ] Jams
- [ ] Maple syrup
- [ ] Pesto
- [ ] Pickles & preserves
- [ ] Popcorn
- [ ] Prepared foods
- [ ] Salsa
- [ ] Sauces
- [ ] Spirits
- [ ] Sunflower oil
- [ ] Tortillas
- [ ] Vinegars
- [ ] Wild rice

Cheeses of the Month: Mozzarella and Provolone

Profile: While authentic Italian mozzarella is made only with high-fat buffalo milk, virtually all American-made mozzarella comes from pasteurized cow milk. But as with the original, mozzarella curds are dipped into hot water to soften and then kneaded and pulled to develop stretchiness. Because of its springy texture, lovely melting quality, and dulcet milk flavor, fresh mozzarella is the darling of artisan pizza makers, but is consumed even more often in salads and antipasti, uncooked. Regular mozzarella has classically neutral flavor which makes it an ideal palette for other ingredients and flavors. It is the most-produced cheese in Wisconsin. The other major type of pasta filata, or stretched-curd cheese, is Provolone, which is similar to mozzarella, but is aged. Provolone is also a great melter and is often smoked. Cheesemakers seem to have fun with mozzarella and Provolone, shaping the cheese into balls or logs, but also more whimsical forms like melons, pears, and bottles.

Wisconsin-made styles include: Fresh mozzarella, whole milk mozzarella, part-skim mozzarella, Dolce (young Provolone), Picante (aged Provolone), smoked Provolone.

Taste: Fresh mozzarella has a sweet, clean milk flavor, with very faint grassiness. Regular mozzarella is pleasingly bland. Provolone has an herbaceous, earthy savor and a little tang; aged varieties grow tangier and saltier.

Texture/Look: Fresh mozzarella is soft, moist, and giving, with a slightly open-textured interior and a springy chew. Regular mozzarella is smoother, drier, and firmer. Provolone can be found from semi-firm to crumbly.

Go-withs: Grapes, apricots, dried pears, tomatoes, eggplant, basil, olive oil, balsamic vinegar, roasted sweet peppers, marinated mushrooms, cured meats, olives.

What to drink: Beaujolais, Chardonnay, Semillon, port, Sauternes, ice wine, lager, red lager.

Classic dishes: Pizza, antipasti, caprese salad, lasagna, muffaletta sandwich.

Try this: Scatter cooked spinach-filled tortellini around a large platter. Scatter small fresh mozzarella balls around it. Cut Sungold tomatoes in half and add them to the platter. Drizzle all with the best vinegar and oil you can afford. Sprinkle with basil chiffonade, flake sea salt, and freshly ground pepper.

October 2014

Wednesday 1

Thursday 2

Friday 3

Yom Kippur Begins (Sunset)

Saturday 4

Yom Kippur Ends (Sunset)
New Moon

Buttered Quesadillas
Serves: Any number

Quesadillas are one of those go-to preparations that can make a meal or a snack at any time of the day (or night). Kids love them, teens love them, grown-ups love them. Pan-grilled tortillas, fresh veggies, molten cheese—what's not to love? The extra touch with this version comes when you spread a little softened butter on the outside of each tortilla before pan-grilling it. The butter makes them gloriously brown and crispy on the outside and adds an extra savor to the overall taste. Qué rico!

> softened butter
> whole wheat flour tortillas
> grated cheese (sharp Cheddar, sheep milk Gouda, aged Swiss, Pepper jack, etc.)
> finely chopped green onions or grilled yellow onions
> finely chopped green peppers or roasted red peppers
> chopped cilantro
> diced ripe avocados or sautéed zucchini
> bottled or homemade salsa

Heat a large cast-iron griddle or two heavy skillets over a medium flame until quite hot, about 10 minutes.

Meanwhile, spread a very thin layer of softened butter on one side of each tortilla, laying them down on waxed paper, buttered side down.

When the griddle is hot, spread desired amounts of each of the remaining ingredients across one half of the unbuttered side of a tortilla. Then dab a tiny bit of butter at the far edge of the tortilla half that has no filling. Fold tortilla in half over the filling ingredients, pressing it near the dab of butter to help seal it. Place on hot griddle. Continue filling tortillas to fill the griddle.

Cook quesadillas on both sides until cheese is melted and tortilla surface is browned. Serve hot.

October 2014

Sunday	5
Monday	6
Tuesday	7
Wednesday	8
Thursday	9
Friday	10
Saturday	11

Apple Cider Cheddar Fondue

Servings: 8

Fondue is the dish of camaraderie: it has that everyone-around-the-pot way of bringing people together. What's more, you can get it to table in less than twenty minutes. If you don't own a fondue pot, use a heavy saucepan instead, and plan on gently reheating the cheese sauce once or twice during dinner.

This recipe calls for apple cider—be sure to use real cider and not apple juice. (We haven't tried it with hard cider but suspect that would be wonderful.) For another, more classic fondue, substitute aged Swiss or Emmentaler for the Cheddar and dry white wine for the cider, and rub the pot with half of a garlic clove before beginning the recipe.

> 1 pound medium or sharp Cheddar, finely diced
> 4 tablespoons flour
> 2 cups apple cider
> 1 ½ tablespoons fresh lemon juice
> ⅛ teaspoon pepper
> sourdough, whole-grain, or rye bread, cut into 1-inch cubes

Mix cheese and flour in a bowl. Combine apple cider and lemon juice in fondue pot. Bring to very low simmer; do not let it boil. Add a handful of cheese and stir constantly until the cheese is fully melted. Continue to add one handful of cheese at a time, stirring constantly and melting the cheese fully before adding the next handful. After all the cheese is fully melted, stir in pepper.

Serve fondue with a basket of cubed bread and long forks for dipping.

October 2014

| | Sunday | 12 |
| Monday | 13 |
Columbus Day

	Tuesday	14
	Wednesday	15
	Thursday	16
	Friday	17
	Saturday	18

Penne with Tomato, Morel, and Mascarpone Sauce

Servings: 6

A pasta dish that's company-worthy, and then some. Should you happen to have any dried-out Parmesan ends on hand, add a few to the sauce with the tomatoes. It will give the dish extra depth. If dried morels are not available, go for dried porcini instead.

    1 ½ ounces dried morels (about 2 cups), rinsed
    boiling water
    2 tablespoons olive oil
    ¼ cup finely chopped onion
    2 teaspoons minced garlic
    1 canned chipotle pepper in adobo sauce
    1 can (28 ounces) Italian peeled tomatoes in tomato purée
    ⅔ cup mascarpone cheese
    2–3 sage leaves, minced (about 1 tablespoon)
    12 ounces penne rigate (a ridged, tube-shaped pasta)
    1 cup freshly grated Parmesan

Place dried morels in a 2-cup glass measuring cup. Pour in boiling water to reach the 2-cup measure mark. Place a coffee mug atop the mushrooms to immerse them. Let stand until mushrooms are plumped, about 10 minutes. Remove mushrooms and rinse them well under cool water. Squeeze them lightly to remove excess liquid. Cut them in half if they're large. Strain mushroom liquid through cheesecloth into a bowl.

Heat olive oil in a heavy saucepan over a low flame. Add onion and garlic; cook slowly a few moments. Mince the chipotle and stir it into onions. Raise heat to medium-high and add mushrooms. Cook a few moments, stirring often. Add ⅔ cup mushroom liquid and simmer, stirring occasionally, until most of the liquid is reduced.

Coarsely chop the tomatoes; add them plus the sauce from the can to the mushrooms. Bring to simmer and cook, stirring occasionally, 20–30 minutes. Season with salt and pepper. At this point you can turn off the sauce and cool it (or chill it) until about ½ hour before serving.

To finish: Bring large pot of salted water to boil; add penne and cook until barely tender. Meanwhile, return sauce to a simmer. Stir in mascarpone and sage. Taste and add more salt and pepper, if necessary. When pasta is done, drain and place it in a large serving bowl. Add the sauce and about half the Parmesan; toss well. Serve immediately, sprinkling each serving with additional Parmesan.

October 2014

Sunday	**19**
Monday	**20**
Tuesday	**21**
Wednesday	**22**
Thursday	**23**
	New Moon
Friday	**24**
Saturday	**25**
	Islamic New Year

Robert Wills, President, Owner, and Master Cheesemaker

Cedar Grove Cheese and Clock Shadow Creamery
E5904 Mill Road, Plain (Cedar Grove Cheese)
138 West Bruce Street, Milwaukee (Clock Shadow Creamery)
800-200-6020
cedargrovecheese.com

Bob Wills left the world of economics and law to become a cheesemaker. He and his wife Beth bought the well-established Cedar Grove Cheese Company, a plant in existence since 1878, from Beth's parents in 1989. Bob learned how to make cheese from his father-in-law Ferdinand Nachreiner and from Dan Hetzel, senior cheesemaker at the firm with over 50 years of experience. Bob also took classes at the Center for Dairy Research at the University of Wisconsin-Madison, which he says helped add science to the art. He holds Wisconsin Master Cheesemaker certifications for Cheddar and Butterkäse cheeses and oversees production of many other cheese varieties at the plant, including artisanal and organic ones.

In 1993 Cedar Grove was the first company in the U.S. to notify its customers that their cheese production used no milk from cows treated with recombinant bovine growth hormone (rBGH) and, in 2001, that it used no genetically modified organisms. Today, over half of the cheeses the company produces are organic.

Bob is also a cheesemaker's cheesemaker. He has provided mentoring and space in his plant to launch small-scale cheesemakers. Another facility earmarked for similar mentoring is Bob's urban cheese factory, Clock Shadow Creamery, which opened in Milwaukee in 2012.

Where do you get your inspiration to craft a new cheese? From the milk. We consider the animals, what they are eating, the milk quality and freshness. Then we mix in the farmer's background, the target consumer's values and tastes, the available time and space for curing and a dose of curiosity. Then we make a bunch of mistakes until we find something we think does the milk justice.

Why is Wisconsin the only state to offer a Master Cheesemaker program?
A program needs a critical mass of interested and experienced participants to be successful. Few other areas have the concentration of experienced cheesemakers to provide that demand.

What is it about cheese that makes it an ideal ingredient to cook with?
It comes in many forms and, therefore, can provide a wide variety of textures and flavors when added to other foods. Heat changes its flavor in ways that add interest and comfort.

What's your favorite thing about cheese? There is always more to learn.

October 2014

Sunday	**26**
Monday	**27**
Tuesday	**28**
Wednesday	**29**
Thursday	**30**
Friday	**31**
Halloween	
Saturday	1

Joyce Penterman (Holland's Family Cheese), second generation cheesemaker.
Photo courtesy of Holland's Family Cheese.

Great cheese is no joke…or is it? Prepare to chuckle (or groan):

KEN HEIMAN, NASONVILLE DAIRY:
   "What do you call cheese that doesn't belong to you? Nacho cheese."

JOE BURNS, BRUNKOW CHEESE:
   "My nine year old son Henry got me on this joke:
      Henry: Dad, I have a great joke for you about a piece of pizza.
      Me: Okay, Hank, let's hear it.
      Henry: Nahh, you won't like it, it's too cheesy."

KERRY HENNING, HENNING CHEESE:
   "Cheese makers never leave; they just get in the whey."

VEGETABLES (INCLUDES HYDROPONIC & STORED)

- [ ] Beets
- [ ] Broccoli
- [ ] Brussels sprouts
- [ ] Burdock root
- [ ] Cabbage
- [ ] Carrots
- [ ] Cauliflower
- [ ] Celeriac
- [ ] Chinese cabbage
- [ ] Dried beans
- [ ] Dried chiles
- [ ] Garlic
- [ ] Green tomatoes
- [ ] Kohlrabi
- [ ] Leeks
- [ ] Mushrooms (cultivated)
- [ ] Onions
- [ ] Parsnips
- [ ] Potatoes
- [ ] Pumpkins
- [ ] Rutabaga
- [ ] Shallots
- [ ] Sprouts
- [ ] Sweet potatoes
- [ ] Turnips
- [ ] Winter radishes
- [ ] Winter squash

HERBS & FRESH GREENS

- [ ] Arugula
- [ ] Kale
- [ ] Lettuce
- [ ] Microgreens
- [ ] Oregano
- [ ] Parsley
- [ ] Rosemary
- [ ] Sage
- [ ] Spinach
- [ ] Thyme

FRUITS

- [ ] Apples
- [ ] Frozen fruits
- [ ] Pears

MEATS & FISH

- [ ] Beef
- [ ] Bison
- [ ] Chicken
- [ ] Emu
- [ ] Lamb
- [ ] Pork
- [ ] Rabbit
- [ ] Rainbow trout
- [ ] Sausages
- [ ] Smoked fish
- [ ] Turkey
- [ ] Venison
- [ ] Whitefish

EGGS & DAIRY

- [ ] Butter
- [ ] Buttermilk
- [ ] Cheese (cow, goat & sheep milk)
- [ ] Cottage cheese
- [ ] Cream
- [ ] Curds
- [ ] Dips
- [ ] Eggs
- [ ] Ice cream
- [ ] Milk
- [ ] Yogurt

OTHER

- [ ] Baked goods
- [ ] Beer
- [ ] Breads
- [ ] Cider
- [ ] Flours
- [ ] Hickory nuts
- [ ] Honey
- [ ] Horseradish
- [ ] Jams
- [ ] Maple syrup
- [ ] Pesto
- [ ] Pickles & preserves
- [ ] Popcorn
- [ ] Prepared foods
- [ ] Salsa
- [ ] Sauces
- [ ] Spirits
- [ ] Sunflower oil
- [ ] Tortillas
- [ ] Vinegars
- [ ] Wild rice

Cheese of the Month: Swiss

Profile: Worldwide, there are a great many Swiss varieties, including prestigious Emmentaler from Switzerland (where Swiss cheese originated) and the higher fat Gruyère that is often associated with French cooking. Most American Swiss is rindless and milder than its European-made cousins; made from pasteurized cow milk,  it is aged at least sixty days. Not all Swiss cheese has holes, and not all cheese with holes is Swiss cheese. The holes (properly known as "eyes") form when the cheese begins to ferment, releasing carbon dioxide and forming bubbles. Cheesemakers determine the eye size by altering acidity, temperature, and maturing time. Green County, with its rolling hills and Swiss heritage, is the state's epicenter of Swiss cheese.

Wisconsin-made styles include: Swiss, Baby Swiss, Gruyère, Emmentaler, Alpine Lace, Raclette.

Taste: Nuts, butter, woods. Baby Swiss and other young varieties can be notably sweet, while aged varieties can carry a hint of apple, fresh hay, or aged leather. Available smoked.

Texture: Ranges from soft to extra-hard, but most Swiss-type cheeses are medium-hard and have a smooth pliability. Nearly all Swiss styles melt well.

Go-withs: Dark rye bread, ham, caramelized onions, mushrooms, asparagus, eggplant, celeriac, cauliflower, mushrooms, leeks, spinach, winter squash, pears, eggs, potatoes, almonds, thyme, nutmeg, rosemary.

What to drink: Gewürztraminer, Riesling, Chardonnay, Malbec, Cabernet Sauvignon, doppelbock, dark ale, lambic, spiced whitbier, apple or pear cider.

Classic dishes: Rarebit, French onion soup, omelettes, röesti, fondue, quiche.

Try this: Baked Reuben Soup with Swiss Cheese and Rye – Sauté onions and caraway seeds in butter. Add sauerkraut, shredded corned beef, chopped pickles, dark beer, and beef or chicken stock. Simmer and then let it cool completely to develop flavor. Reheat soup, ladle into ovenproof bowls, and top with rye toast and a mound of grated Swiss. Bake at 450 degrees until bubbly and brown-spotted.

November 2014

Sunday	26
Monday	27
Tuesday	28
Wednesday	29
Thursday	30
Friday	31
Saturday	1

Coyote Chicken Co-Jack Wraps

Servings: 4–6

The November chill is on, but if you're like us, you still have a few end-of-the-season tomatoes ripening in your pantry. And if you're lucky like us, you also know an indoor farmers' market vendor who sells fresh tortillas. So get out a baking dish, turn on the oven, and bring some Southwestern warmth into your life with these festive wraps. The recipe is from Jane Patton Walsh of Madison, who won an Honorable Mention for it in the 2000 Food for Thought Recipe Contest, sponsored by the REAP Food Group.

> 2 tablespoons butter
> 1 cup chopped onions
> 1 cup sliced fresh mushrooms
> 2 medium garlic cloves, minced
> 2 cups chopped, cooked chicken meat (leftover grilled or roasted
>   chicken is best)
> ¾ cup pine nuts
> 1 cup seeded, chopped tomatoes
> ½ cup chopped cilantro
> ¼ teaspoon sea salt
> 2 cups grated Co-Jack cheese, divided
> 8 flour tortillas (6- or 8-inch size)
> homemade or bottled tomato or mango salsa

Heat butter in large skillet over a medium flame. Add onions, mushrooms, and garlic and sauté, stirring often, until onions are translucent, about 4 minutes.

Heat oven to 375 degrees. Oil a 9-by-13-inch baking pan. Combine chicken, pine nuts, tomatoes, cilantro, sea salt, and 1 cup of the grated cheese in bowl.

Place a little of the chicken mixture and a little of the mushroom mixture in the center of a tortilla; roll up and place in pan with seam side down. Repeat with remaining tortillas and most of both the mixtures. Scatter remaining mixtures over rolled tortillas. Sprinkle with remaining cheese. Sprinkle with a little water. Cover with aluminum foil, and bake about 35 minutes.

Serve hot with homemade or bottled salsa of choice.

November 2014

Sunday	2

REAP's Pie Palooza
Daylight Saving Time Ends

Monday	3

Tuesday	4

Election Day

Wednesday	5

Thursday	6

Friday	7

Saturday	8

Pleasant Ridge Reserve and Rosemary Scalloped Potatoes
Serves: 8–10

Made from the milk of a single herd that grazes near Dodgeville, the cheese called Pleasant Ridge Reserve has an unmatched claim to fame: it has won Best in Show not once, not twice, but three times in the prestigious American Cheese Society competition. On top of that, it took top honors in a U.S. Cheese Championship. For years this cave-aged, Gruyère-like specialty was the only variety produced by Uplands Cheese company, but now fans can enjoy a second treat from the farmstead operation—the luscious Rush Creek Reserve. Another raw milk, washed-rind cheese, Rush Creek Reserve is a soft-textured, spruce-wrapped marvel with woodsy flavor and incredible richness.

This recipe calls for Russets, which have a high starch content that will absorb the liquid and create a creamy sauce. Do not use red-skinned or other boiling, or waxy, potatoes. (And yes, you can make this dish with other cheeses—any good-melting, full-flavored Wisconsin cheese will do the trick, even if it isn't world famous.)

    softened butter for the baking dish
    2 cups heavy cream
    1 ¾ cups whole milk
    1 ½ teaspoons sea salt
    freshly grated pepper to taste
    3 pounds Russet potatoes
    1 ½ tablespoons minced fresh rosemary, divided
    6 ounces Pleasant Ridge Reserve, shredded

Heat oven to 375 degrees; generously butter a large, wide baking dish that is at least 2 ½ inches deep.

Combine heavy cream, milk, salt and pepper in a large pot. Peel and slice the potatoes very thinly, placing them in the cream mixture as you go. Bring potato-cream mixture to simmer over a medium flame; simmer 5–7 minutes. Stir in most of the rosemary.

Transfer potato mixture to the baking dish. Place a large piece of aluminum foil on rack in oven and set baking dish on the foil (the foil will catch drips should the potatoes bubble over a bit). Bake 20 minutes. Spread shredded cheese over the potatoes and continue to bake until they are fully tender and the dish looks thickened, bubbly, and brown, another 20–30 minutes. Remove from oven, sprinkle with remaining rosemary, and let stand at 15–20 minutes before serving.

November 2014

Sunday	**9**
Monday	**10**
Tuesday *Veteran's Day*	**11**
Wednesday	**12**
Thursday	**13**
Friday	**14**
Saturday	**15**

Kale and Swiss Cheese Pie with Pasta Crust

Servings: 8–10

Who ever heard of a pasta crust? We have, and we like it a great deal. Bound with eggs and baked until set, it is the base for a savory pie made with Swiss cheese and frost-sweetened kale. Feel free to add a little bacon or ham for extra heartiness.

2 tablespoons butter, plus a little for the baking dish
8 ounces thin spaghetti, broken into thirds
¼ cup shredded Parmesan
6 large eggs, divided
1 tablespoon olive oil
¼ cup finely chopped shallots
8–10 ounces fresh kale, stems removed
4 ounces Swiss, shredded
½ cup chopped ham or cooked bacon (optional)
⅓ cup milk
⅛–¼ teaspoon red pepper flakes
salt and pepper

Heat oven to 350 degrees. Butter an 8-by-12-inch baking dish. Cook spaghetti in lots of salted boiling water until tender. Drain, place in a bowl, and stir in butter and Parmesan. Let pasta cool down for a few minutes. Beat 2 of the eggs and stir them into the pasta. Place the mixture in the baking dish and press it all over to even it out. Cover with foil and bake 13 minutes. Set the crust aside.

Heat olive oil in a large skillet or pot, add shallots, and cook until tender. Add kale and stir over heat until wilted. Remove from heat, let kale cool down a bit, and then coarsely chop it. Beat remaining 4 eggs and add to kale along with the cheese, ham or bacon (if you're using either one), milk, and red pepper flakes. Add salt and pepper to taste and mix well.

Spread filling over crust. Cover again with foil. Bake 40 minutes. Uncover and bake another 5 minutes. Let the pie stand 10 minutes before serving.

November 2014

Sunday	**16**
Monday	**17**
Tuesday	**18**
Wednesday	**19**
Thursday	**20**
Friday	**21**
Saturday	**22**

New Moon

Photo by Becca Dilley

Bruce Workman, Master Cheesemaker and Owner

Edelweiss Creamery
W6117 County Highway C, Monticello
608-938-4094
edelweisscreamery.com

The star attraction at Edelweiss Creamery is Master Cheesemaker Bruce Workman's 180-pound wheels of Emmentaler cheese crafted from grass-based, raw milk the traditional Swiss way. To make these old-world beauties, Bruce equipped the cheese factory he renovated (the former Prima Kase plant in Monticello in 2003) with a cheese press and copper-lined vat purchased from a master cheesemaking school in Switzerland. The cheese's characteristic nutty flavor is attributed to a reaction of the milk with the vat's copper lining. Bruce cures his Emmentaler up to a year, ensuring that the flavor and texture equals the celebrated European original.

Bruce also makes Butterkäse, a traditional German Brick cheese; Muenster, a mild melting cheese; Havarti, a higher-fat Danish cheese, plain or flavored with onion, dill, or chile peppers; a reduced-fat, low-sodium Baby Swiss; and Gouda, another Danish cheese.

How did you become a cheesemaker? I always enjoyed being in the kitchen with my mother, a home economics teacher, when she prepared food. She taught me that patience is needed to create superb recipes. Now I use really big pots when [cooking up] artisanal cheeses. My first experience in cheesemaking came when I visited the Roelli Cheese factory in Darlington at the age of 12. I was inspired by the transition of milk to cheese and the dedication of the cheesemakers. It wasn't until we moved to Monticello in 1971 that [this] city boy found a job working at the North Side Cheese Co-op—under cheesemakers Fritz Minder and Bob Durtschi. I loved the work, and in 1973 I became a Wisconsin licensed cheesemaker. In 1996 I enrolled in the Master Cheesemaker Program. I have been through the program five times and currently hold the most certificates (nine) as a Wisconsin Master Cheesemaker.

What is it about cheese that makes it an ideal ingredient to cook with? The availability to select a cheese with the functions you need is what I really like. High-flavor, harder cheeses for adding that zing to your dish, or the semi-soft cheese for melting.

Where do you get your inspiration to craft a new cheese? I listen to what the consumers are talking about. Some want a new flavor added to the cheese. Others are looking for a certain flavor profile from one type of cheese but want the melting capabilities of another type. Sometimes it is just bringing up old cheese types that seem to be trendy for a while but always come full circle.

November 2014

Sunday	**23**
Monday	**24**
Tuesday	**25**
Wednesday	**26**
Thursday — *Thanksgiving Day*	**27**
Friday	**28**
Saturday/Sunday	**29/30**

Layered beets and mascarpone.
Photo courtesy of Crave Brothers Farmstead Cheese. (Digitally altered)

Why is Wisconsin such an out-and-out amazing place for cheese? Sue Merckx, who manages retail marketing for Sartori Cheese, explains: "Wisconsin is home to a lot of true, artisan cheese producers. The soil and natural elements of climate converge to create remarkable growing conditions. This allows [us] to grow high quality forages, which allows the cows to produce premium-grade milk. The milk is high in component value, protein, butterfat, and other solids, which in turn makes for a very distinctive cheese.

"Wisconsin also has a long tradition of having the highest quality standards in the nation. Additionally our infrastructure is outstanding. Not only are cheesemakers served by excellent equipment manufacturers, suppliers, etc., but Wisconsin dairy farmers are supported by the best large animal veterinarians, nutritionists, and so on."

VEGETABLES (INCLUDES HYDROPONIC & STORED)

- [ ] Beets
- [ ] Brussels sprouts
- [ ] Burdock root
- [ ] Cabbage
- [ ] Carrots
- [ ] Cauliflower
- [ ] Celeriac
- [ ] Chinese cabbage
- [ ] Dried beans
- [ ] Dried chiles
- [ ] Garlic
- [ ] Kohlrabi
- [ ] Leeks
- [ ] Mushrooms (cultivated)
- [ ] Onions
- [ ] Parsnips
- [ ] Potatoes
- [ ] Pumpkins
- [ ] Rutabaga
- [ ] Shallots
- [ ] Sprouts
- [ ] Sweet potatoes
- [ ] Turnips
- [ ] Winter radishes
- [ ] Winter squash

HERBS & FRESH GREENS

- [ ] Arugula
- [ ] Kale
- [ ] Lettuce
- [ ] Microgreens
- [ ] Oregano
- [ ] Parsley
- [ ] Rosemary
- [ ] Sage
- [ ] Spinach
- [ ] Thyme

FRUITS

- [ ] Apples
- [ ] Frozen fruits
- [ ] Pears

MEATS & FISH

- [ ] Beef
- [ ] Bison
- [ ] Chicken
- [ ] Emu
- [ ] Lamb
- [ ] Panfish
- [ ] Pork
- [ ] Rabbit
- [ ] Rainbow trout
- [ ] Sausages
- [ ] Smoked fish
- [ ] Turkey
- [ ] Venison
- [ ] Whitefish

EGGS AND DAIRY

- [ ] Butter
- [ ] Buttermilk
- [ ] Cheese (cow, goat & sheep milk)
- [ ] Cottage cheese
- [ ] Curds
- [ ] Dips
- [ ] Eggs
- [ ] Ice cream
- [ ] Milk and cream
- [ ] Yogurt

OTHER

- [ ] Baked goods
- [ ] Beer
- [ ] Breads
- [ ] Cider
- [ ] Flours
- [ ] Hickory nuts
- [ ] Honey
- [ ] Horseradish
- [ ] Jams
- [ ] Maple syrup
- [ ] Pesto
- [ ] Pickles & preserves
- [ ] Popcorn
- [ ] Salsa
- [ ] Sauces
- [ ] Spirits
- [ ] Sunflower oil
- [ ] Tortillas
- [ ] Vinegars
- [ ] Wild rice

Cheese of the Month: Italian-Style Hard Cheese

Profile: Hard, low-moisture Italian-style varieties (often called granas) are old cheeses, both in terms of their history and their shelf life. Parmesan was developed by Italian monks during the twelfth century, Asiago around the year 1000. Aging is essential to their character and consistency—they gain flavor, richness, and firmness over months of proper storage. Used mostly as a grating or finishing cheese, many varieties also make excellent table cheeses. Typically produced from low-fat milk and used sparingly because of the intense flavor, they can also be a boon for people watching their calories. Don't let anyone tell you that the only really good grana is Parmigiano-Reggiano. It is indeed one of the world's great cheeses, but there are critically acclaimed Wisconsin varieties that are right up there with Italy's best.

Wisconsin-made styles/brands include: Parmesan, Asiago, Pecorino, Romano, Sartori's BellaVitano and SarVecchio Parmesan, BelGioioso's American Grana.

Taste: Toasted nuts, butter, umami. Lightly piquant to assertively sharp. Typically there is some sweetness, and some varieties are markedly fruity. Pecorino, made from sheep milk, has a characteristic tanginess.

Texture: Very firm to hard. Typically granular—some of the very best varieties even bear a delightful crystalline crunch—but some near-smooth ones exist too, especially when sold young.

Go-withs: Tomatoes, eggplant, fennel, mushrooms, winter squash, asparagus, olive oil, peppercorns, rosemary, filberts, walnuts, dried fruit, basil, crusty bread, cured sausage.

What to drink: Prosecco, Chardonnay, Sauvignon Blanc, Cabernet Sauvignon, Nebbiolo, malty dark lager, nut brown ale, India pale ale, Belgian-style saison, port, sherry, hard cider.

Classic dishes: Eggplant Parmesan, risotto, chicken Parmesan, pesto, Caesar salad.

Try this: Parmesan Lace – Combine 4 ounces shredded Parmesan and 1–2 teaspoons minced fresh thyme. Heat a non-stick skillet over medium-low flame. Sprinkle 1 ½ tablespoons Parmesan mixture into a 3 ½-inch circle in the pan. Cook until a bit melted and beginning to color, 1–2 minutes. Work a spatula underneath to loosen, then flip and cook the other side for a moment. Transfer to a paper towel. The cheese will firm up as it cools. Repeat procedure with remaining cheese.

December 2014

Saturday/Sunday 29/30

Monday 1

Tuesday 2

Wednesday 3

Thursday 4

Friday 5

Saturday 6

Italian Egg Drop and Cheese Soup

Servings: 4 small or 2 big bowls

You don't need scads of cheese to make a culinary statement. Indeed, it's amazing how a little cheese can go a long way, as is the case with this stunning little soup. The quality of cheese really matters, of course, as does the richness of the broth, so avoid pre-grated Parmesan and canned chicken stock here. Go for real-deal aged Parmesan or Romano, and some of that frozen broth you made from leftover poultry bones last month. And if you're an indoor gardener, snip fresh herbs from your windowsill pots into it, too.

The flavors in this soup are provocative enough for it to be served as the first course of a special-occasion dinner. But it's so darn easy to make, you might as well add it to your everyday repertoire.

4 cups chicken broth
2 large eggs
4 tablespoons coarsely grated hard Italian-style cheese, such as
   Parmesan or Romano
2 tablespoons panko-style bread crumbs
3 tablespoons chopped fresh parsley
1 teaspoon chopped fresh rosemary
1 ½ teaspoons minced garlic, mashed to a paste with flat of knife
salt and pepper
grated lemon zest

Place chicken broth in a saucepan over a medium flame; bring to simmer. Mix eggs, cheese, bread crumbs, parsley, rosemary, and garlic in a bowl. Whisk this mixture into the simmering stock; cook, whisking, for 1 minute.

Season with salt and pepper to taste. Serve each bowl with a sprinkling of grated lemon zest on top.

December 2014

Sunday	7
Monday	8
Tuesday	9
Wednesday	10
Thursday	11
Friday	12
Saturday	13

Feta Tart

Servings: 8–10

When a cook writes, "This is hugely good; I can't wait to make it again" next to a recipe, you know you're onto something special. That's what we saw when we were sorting through recipes for inclusion in this collection, so of course we had to share. This one's adapted from a recipe in Saveur magazine and seems to fit into any meal of the day or season of the year. During the cold months, serve it with cocktails before a holiday feast, or with a bowl of tomato-dill soup for a mid-week supper.

1 ¼ cups flour
¼ teaspoon salt
¼ teaspoon baking powder
5 tablespoons extra-virgin olive oil, divided
2 teaspoons bourbon
1 egg
6 ounces crumbled feta
2 tablespoons unsalted butter, cut into bits

Place a heavy 18-by-13-inch baking pan with sides in cold oven. (Don't use a lightweight pan—it will warp in the hot oven.) Heat oven to 500 degrees.

While oven is heating, whisk the flour, salt, and baking powder in a bowl. Using another bowl, combine 2 tablespoons of the olive oil, the bourbon, egg, and 1 cup of water. Whisk well. Pour wet mixture into dry and whisk until smooth.

After the baking pan has heated at 500 degrees for at least 10 minutes, remove it from the oven, and brush it with the remaining 3 tablespoons olive oil. Spread the batter evenly in the pan (it doesn't have to reach all the way to the sides). Sprinkle cheese over top. Scatter the butter bits over cheese. Bake until cheese is brown-tipped and crust has set, 20–25 minutes. Serve hot or warm.

December 2014

Sunday	14
Monday	15
Tuesday	16
Hanukkah Begins (Sunset)	
Wednesday	17
Thursday	18
Friday	19
Saturday	20

Sheep Milk Cheese Gnocchi with Hickory Nuts, Garlic, and Sage Butter Sauce

Servings: 2

Once you taste this recipe, you won't be surprised that it took first place in its category in the 2005 Food for Thought Recipe Contest, sponsored by the REAP Food Group. The creator is Brian Garthwaite, who wrote, "A simple sage butter sauce shows off the gnocchi, but they're also great with a rich tomato sauce, in soups, as a side dish for rich meats, etc."

Gnocchi:
½ cup fresh sheep milk Brebis cheese
1 egg
1 tablespoon grated Parmesan
pinch each of freshly grated nutmeg and salt
1 cup white or whole wheat flour (or use a combination)
Sauce:
2 tablespoons butter
⅓ cup hickory nuts
1 clove garlic, minced
small handful fresh sage, sliced into very fine ribbons
freshly grated Parmesan
freshly grated black pepper

Bring a large pot of salted water to boil. Meanwhile, mix brebis, egg, Parmesan, nutmeg, and salt in small bowl until smooth. Stir in the flour and mix until combined. Turn mixture onto a floured surface and knead until it comes together and there are no more loose pockets of flour. Use your hands to roll out the dough into a series of long, skinny "ropes" that are ½ to ¾ inch in diameter. They should still be fragile enough to break apart at points. Cut into 1-inch lengths. Then, using the back of a fork, flick each piece of dough along the tines to mark it with lines, and make a divot in the back of each with your thumb. Cook the gnocchi in the boiling water until they float to the top. Give them another 30 seconds and then drain them.

For the sauce, melt the butter in a large skillet over medium-high heat. When it stops bubbling, add the hickory nuts and garlic, and cook about 1 minute. Add the gnocchi and allow it to get a little crispy and brown on both sides. Shortly before it's done, sprinkle sage over the top. Serve immediately with freshly grated Parmesan and black pepper.

December   2014

Sunday **21**

First Day of Winter

Monday **22**

New Moon

Tuesday **23**

Wednesday **24**

Last Day of Hanukkah
Christmas Eve

Thursday **25**

Christmas Day

Friday **26**

Kwanzaa Begins

Saturday **27**

Gianni Toffolon, Master Cheesemaker

BelGioioso Cheese, Inc.
4200 Main Street, Green Bay
920-863-2123
belgioioso.com

Gianni Toffolon moved to Wisconsin from Cremona, Italy, in 1979, when the Italian cheesemaker for whom he was working decided to expand to the United States. Now, the award-winning Master Cheesemaker is perfecting his craft at BelGioioso Cheese. Gianni is certified as a Master Cheesemaker in Parmesan and Fontina. His passion for Parmesan stems from over 30 years of experience producing it. "With Parmesan, I learned from other experienced cheesemakers and then put my own personal touch on it. With Fontina, I visited traditional cheese factories in Italy and learned the cheesemaking techniques for this variety, which helped me create Fontina in the United States," Gianni said. He credits much of his knowledge to the Italian cheesemakers, and his work mirrors their teachings. He became involved with the Wisconsin Master Cheesemaker Program so he could learn the science behind the process of cheesemaking, which he felt would make him a better cheesemaker.

Please describe your company. BelGioioso Cheese is a family-owned and -operated company specializing in artisan Italian cheesemaking. Using natural ingredients and fresh, local Wisconsin milk, Master Cheesemakers hand-craft a full line of exceptional cheeses guided by a commitment to quality and a respect for tradition. At BelGioioso, every cheese is a specialty.

Do you make any types of Italian cheese that aren't well-known in the U.S.?

We make several small-batch artisan cheeses that are not domestically well-known, such as Crescenza-Stracchino. This is my personal favorite cheese. Crescenza-Stracchino is made in petite 3½-pound wheels from whole, pasteurized cow milk. This fresh, rindless cheese has a mild flavor with a touch of tartness that tantalizes and makes you want more—one taste is never enough! Its soft texture and creamy consistency allows the cheese to spread and melt wonderfully. Enjoy Crescenza spread on warm crusty Italian bread.

Cheese curds are Wisconsin's biggest seller. If you were going to come up with the next best seller, what would it be?

Fresh Mozzarella Snacking Cheese. The fresh, milky flavor consumers have grown to love is now offered in a convenient one-ounce sized snack portion. These individual packages of America's favorite fresh mozzarella are the ultimate grab-and-go snack at only 70 calories each. Pack them for lunch or eat them as a healthy snack.

December 2014

Sunday **28**

Monday **29**

Tuesday **30**

Wednesday **31**

New Year's Eve

Thursday 1

Kwanzaa Ends

Friday 2

Saturday 3

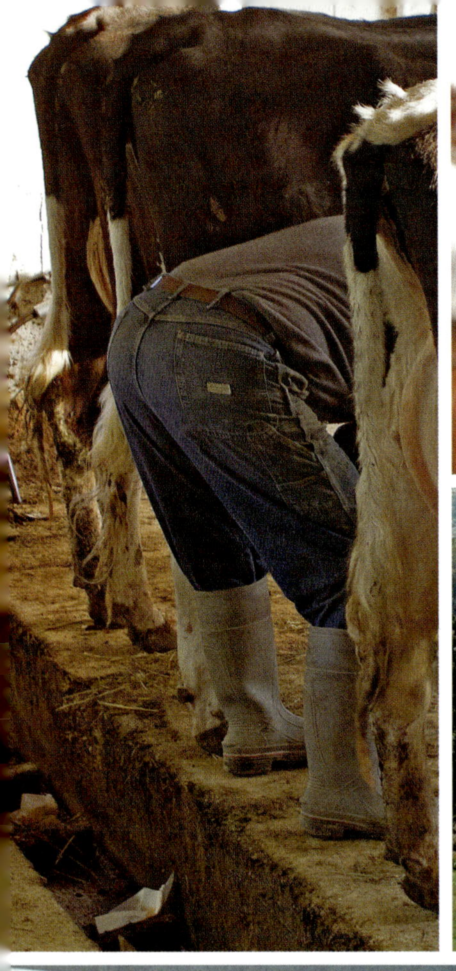

Photo credits - Top left; Bill Lubing. Top right: Becca Dilley. Middle and lower left; ©Wisconsin Milk Marketing Board, Inc. Lower right: Rick Mooney.

Glossary of Cheese Varieties and Cheese Terms

(Excerpted and adapted from eatwisconsincheese.com)

Acid   A descriptive term for cheese with a pleasant tang and sour-ish flavor due to a concentration of acid. By contrast, a cheese with a sharp or biting, sour taste indicates an excessive concentration of acid, which is a defect.

Affinage   The careful practice of ripening or aging cheese, using procedures such as washing (with a flavoring solution) or flipping, to help it reach maximum delectability. See also Finish.

Aged   Generally describes a cheese that has been cured longer than six months. Aged cheeses are characterized as having more pronounced and fuller, sometimes sharper flavors than medium-aged or current-aged cheeses.

Aging   Often referred to as curing or ripening, aging is the process of holding cheeses in carefully controlled environments to allow the development of microorganisms that usually accentuate the basic cheese flavors. See also Curing and Ripening.

American   A descriptive term used to identify the group of American-type cheeses which includes Cheddar, Colby, granular or stirred-curd, and washed or soaked-curd cheeses.

Ammoniated   A term describing cheese that either smells or tastes of ammonia as a result of being overripe or mishandled. A hint of ammonia is not objectionable, but heavy ammoniation is.

American Grana®   A pale-yellow, Italian-style hard cheese made in the United States with the same craftsmanship and care as the Italian original, this special reserve Parmesan is aged over 18 months and has a complex, nutty flavor and crumbly texture.

Annatto   A natural vegetable dye used to give many cheese varieties, especially the Cheddars, a yellow-orange hue. Annatto is odorless, tasteless, and is not a preservative.

Artisan   A term describing cheese made in small batches, often with milk from a limited number of farms, and having unique texture or taste profiles developed in small, sealed production or by specialized producers.

Asadero   A rich Mexican-style melting cheese used for cooking. Creamy, off-white in color, smooth in texture and slightly tangy but mild in flavor. See also Cheese of the Month, page 114.

Asiago   A hard-style cheese that originated in Italy, Asiago in Wisconsin is usually aged to develop sharper flavors and a firm to hard, granular texture. The flavor is buttery and nutty, similar to Parmesan. See also Cheese of the Month, page 150.

Astringent   A term descriptive of a harsh taste with a puckery, almost medicinal quality.

Baby Swiss   Pale yellow in color, soft and silky in texture and mild, buttery, and slightly sweet in flavor, Baby Swiss has the characteristic eyes of other Swiss cheeses. Wisconsin cheesemakers traditionally produce Baby Swiss from whole milk, unlike traditional Swiss cheesemakers, who make it from partially skimmed milk. Whole milk gives Baby Swiss a creamier texture and a more buttery, slightly sweet flavor, which makes it ideal for melting. See also Cheese of the Month, page 138.

Bandaged   Cheddar cheeses wrapped in cheesecloth and dipped in wax. Prior to vapor-barrier film, this method provided the only way to wrap Cheddar cheese for storage and shipping.

Barny or Barnyardy   A descriptive term referring to strong farm-related aromas. Sometimes also called cowy. This characterization does not always indicate a negative quality.

Bitter   An unpleasant, biting flavor—usually an aftertaste. A bitter aftertaste is sometimes associated with variations in manufacturing and curing or aging procedures. Bitterness is often confused with astringency. True bitterness is a sensation that is typified by the aftertaste of grapefruit peel.

Bleu The French word for blue that is used in reference to blue-veined cheese varieties. See also Blue or Blue-veined.

Bloomy Rind A descriptive term for an edible cheese rind (crust) that is covered with a harmless, flavor-producing growth of white Penicillium mold. The bloomy rind is formed by spraying the cheese surface with spores of Penicillium candidum mold before curing. Bloomy-rind cheeses, such as Brie, Camembert, and some chèvres, are classified as soft-ripened.

Blue or Blue-veined Term for cheese varieties that develop blue or green streaks of harmless, flavor-producing mold throughout the interior. Generally, veining gives cheese an assertive and piquant flavor. See also Cheese of the Month, page 66.

Body Represents the physical attributes of cheese when touched, handled, cut, or eaten. The body may feel rubbery, firm, elastic, soft, resilient, yielding, supple, oily, etc. When rolled between the fingers or cut, it may appear waxy or crumbly. Its mouthfeel may be grainy or creamy. A cheese also may be felt to determine its condition of ripeness.

Brick A Wisconsin washed-rind original that is ivory to creamy yellow in color and has a smooth, open texture. The flavor is mild and nutty when young and grows pungent and tangy when the cheese is aged. See also Cheese of the Month, page 16.

Brie The bloomy rind on Brie results from a white mold applied to the surface. The mold produces enzymes which ripen the cheese from the outside in and occurs in just a matter of weeks, giving this cheese a rich, earthy, mushroomy flavor that changes from mild when young to pungent with age. Available plain and flavored, Brie has a soft and creamy interior with a snowy-white edible rind. See also Cheese of the Month, page 78.

Brining A step in the manufacture of some cheese varieties where the whole cheese is floated briefly in a brine solution. Brining is common in the production of mozzarella, Provolone, Swiss, Parmesan and Romano cheeses.

Brushed During the curing process, washed-rind cheese varieties are brushed with liquids such as brine, beer, wine, or brandy to maintain a moist rind and impart distinctive, earthy flavors. Parmesan and other hard cheeses may be brushed or rubbed with a vegetable oil.

bST/Bovine Somatotropin A naturally occurring protein hormone from the pituitary gland of cattle that affects the amount of milk produced by dairy cows. See also rBGH.

Bulk Cheese Cheese in its original manufactured form, such as a 40-pound block of Cheddar.

Burrata A rich, soft, creamy cheese, with an almost buttery texture and sweet, milky flavor, Burrata is hand-formed into 8-ounce balls and packaged in water for an extended shelf life.

Butterfat (Fat, Milkfat) The amount of butterfat/fat in any cheese. Fat content is determined by analyzing the fat in the dry matter of cheese. The fat is expressed as a percentage of the entire dry matter. See also Dry Matter.

Butterkäse A very mild, creamy, pale-yellow cheese that originated in Germany. Although it contains no butter, it has a butter-like texture and is complementary to most foods.

Buttermilk The liquid which remains after churning butter from cultured cream. The liquid remaining after churning sweet cream is sweet cream buttermilk. Also a cultured skim milk.

Buttery A descriptive term for cheese with a high fat content, such as the double and triple creams, or cheese with a sweet flavor and creamy texture reminiscent of butter.

Camembert A bloomy rind cheese made very similarly to Brie. It has a soft, creamy interior and snowy-white, edible rind and a rich, earthy mushroom flavor that becomes more pungent with age. See also Cheese of the Month, page 78.

Casein The principal protein in milk. During the cheesemaking process, casein solidifies, curdles, or coagulates into cheese through the action of rennet.

Chalky (Color) A desirable attribute referring to the true white color or smooth, fine-grained texture of older chèvres and young Brie. However, a chalky appearance on the surface is undesirable in many cheese varieties, such as Cheddar.

Chalky (Mouthfeel) A dry, grainy sensation usually caused by insoluble proteins. Sometimes described as powdery. Generally not a desirable characteristic.

Cheddar A term used to classify cheeses that share characteristics exemplified by Cheddar that may include the process of manufacture, consistency, texture, odor, or flavor. Usually golden but also available in white (which has no dye added), Cheddar has a rich, nutty flavor that becomes increasingly sharp with age, and a smooth, firm texture that becomes more granular and crumbly with age. See also Cheese of the Month, page 40.

Cheddaring The process used in making Cheddar whereby piles of small curds, which have been separated from the whey, are pressed together and cut into slabs. The slabs are then repeatedly turned over and stacked to help drain additional whey and aid in the development of the proper acidity (pH) and body of the cheese. These slabs are then cut or milled into curds and placed in the cheese forms and pressed.

Cheese Curds Collected before cheese (usually Cheddar) has coagulated, cheese curds are small, randomly shaped chunks of premature cheese. They have a mild, salty flavor and a moist, springy bite. When very fresh, they squeak against one's teeth.

Cheesemonger An American term for a cheese sales person.

Chèvres The plural form of the French word for goat originally used to classify all goat cheeses produced in France, but now commonly refers to all goat cheeses, regardless of their origin.

Clean (1) A descriptive term for cheese that is free of unpleasant aromas and off flavors. (2) A lack of lingering aftertaste when eating cheese (i.e., a clean finish).

Close A descriptive term for cheese with a smooth, tight texture, such as Cheddar. A close texture contains few, if any, mechanical holes. A cheese with small holes, like Colby, is characterized as open. See also Open.

Coagulation See Curdling.

Colby A Wisconsin original created in Colby, Wisconsin, in the late 1800s, Colby is a close relative of Cheddar but with a softer, more elastic texture, higher moisture content, a milder flavor, and an open lacy texture. See also Cheese of the Month, page 90.

Cold Pack (Club Cheese) A blend made from different batches of cheeses of the same variety, or two or more varieties of mild and sharp natural cheese that have been ground (comminuted). Unlike processed cheese, Cold Pack is not heat-treated nor cooked at the time of packaging.

Cold Performance Addresses how the cheese responds to mechanical manipulation, such as cubing, shredding, and grating.

Cooked (1) Nearly all milk is heated or warmed to some degree during cheesemaking; however, the term cooked is reserved for those varieties whose curd is heated in order to regulate moisture content and degree of hardness. (2) As a tasting term, cooked refers to a flavor aroma associated with the use of over-pasteurized milk.

Cotija The Parmesan of Mexico, ivory-colored Cotija is firm, crumbly, and salty, and is used as an ingredient, a seasoning, and a garnish. Because of the popularity of Mexican foods in the U.S., skilled Wisconsin cheesemakers have been producing this ethnic cheese in America's Dairyland for many years. See also Cheese of the Month, page 114.

Cream Cheese An American original, cream cheese became popular around 1880. Creamy-white, smooth, and very soft, it is rich, nutty, and slightly sweet. Also available in a lower-fat variety called Neufchatel.

Creams (Single, Double or Triple) A classification of cheese derived from the butterfat content on a dry matter basis. Single creams contain at least 50 percent butterfat in the cheese solids (dry matter); double creams contain at least 60 percent butterfat; and triple creams contain 72 percent or more butterfat.

Creamy   (1) A descriptive term for cheese texture or taste. Creamy texture is soft, spreadable, and in some cases, runny. Creamy flavors are characterized as rich and are associated with cream-enriched cheeses, such as double or triple creams. (2) May also refer to color.

Crumbles   Some cheeses are impossible to shred or grate but will break apart into small sprinkle size portions. Crumbles is a style for cheeses such as feta and blue, to use on salads, pizzas, etc.

Culture (Starter)   A culture normally consists of varying percentages of lactic acid, bacterial or mold spores, enzymes, or other micro-organisms and natural chemicals. Starter cultures speed and control the process of curdling milk during cheesemaking in part by converting lactose to lactic acid. They also lend unique flavor characteristics to the cheese.

Curd   Curdled milk from which cheese is made.

Curdling (Coagulation)   A step in cheese manufacture when casein, the major protein in milk, is clotted by the action of rennet or acids.

Curing   The method, conditions, and treatment from manufacturing to market, such as temperature, humidity, and sanitation, that assist in giving the final cheese product the distinction of its variety. See also Aging and Ripening.

Degree of Hardness   Categorizing cheese by the degree of hardness is the most universal method used. Federal Standards of Identity dictate the tolerances of moisture and milkfat that can be contained in cheese. Since the amount of moisture and fat in cheese significantly controls the properties of the cheese, using degrees of hardness stands on a legal definition.

Double Cream   Term for cheese containing at least 60 percent butterfat in the cheese solids.

Dry Matter   All the components of cheese (solids) excluding moisture (water). Dry matter includes proteins, milkfat, milk sugars, and minerals.

Earthy   A descriptive term for cheese varieties with rustic, hearty flavors and aromas.

Emmentaler   The eyed cheese that originated in the Emme Valley of Switzerland, sometimes referred to as Swiss cheese. See also Swiss and Cheese of the Month, page 138.

Edam   Edam cheese originated in Holland over 800 years ago. This pale-yellow, part-skim milk variety has a light, buttery, nutty flavor and a smooth, firm texture. See also Cheese of the Month, page 28.

Emulsifier   A substance or mixture used in the production of processed cheese to create its smooth body and texture. It is composed of the salts of common food acids.

Eye   A void or hole within cheese caused by the formation of trapped gas as a result of fermentation during the curing process.

Farmers Cheese   Originated on farms throughout the world as a way to use milk left over after skimming the cream for butter. Two main styles evolved: a fresh cheese similar to cottage cheese and a semi-soft version cured for a short time. Wisconsin cheesemakers produce the latter, which is firm enough for cubing or shredding, has ivory to buttery colors with tiny mechanical holes, a smooth, supple texture, and buttery, slightly acidic flavor.

Farmstead   A term describing cheese made on the farm from the milk of that farm.

Fat Content   The amount of butterfat/fat in any cheese. Fat content is determined by analyzing the fat in the dry matter of cheese. The fat is expressed as a percentage of the entire dry matter. In reference to cheese fat, milkfat and butterfat are synonymous.

Feed   A descriptive term for cheese that exhibits an odor or taste that is directly related to the particular feed consumed by a cow or other animal before milking. The aroma or flavor may be unpleasant if the feed was turnips, or intriguing if the feed was apples or mountain clover.

Feta   First made in Greece from sheep or goat milk, chalky white feta now is also made from cow milk. It is packed in brine, which produces a tart and salty flavor and gives the cheese a crumbly, moist texture. See also Cheese of the Month, page 102.

Filled   A descriptive term for cheese from which all butterfat has been removed and in its place a vegetable oil has been used as a substitute. Filled cheese also is referred to as imitation.

Finish   (1) The process of finishing, refining, or curing cheese to desired ripeness. Soft-ripened cheeses are sprayed on the surface with a harmless white mold whose growth helps ripen the cheese. Other finishing methods include washing the rinds of cheeses and the daily turning of cheeses. Temperature and humidity are tightly controlled during the finishing process. (2) The aftertaste of cheese may be described as having a clean finish, bitter finish, sour finish, earthy finish, and so forth.

Flaky   A descriptive term for cheese that breaks into flakes when cut, like Parmesan.

Fontina   A smooth, supple, semi-soft cheese with tiny holes and, in the original Italian style, a mild, earthy, buttery flavor. Excellent as both a table cheese and a cooking cheese. Fontina has been copied often, with the most notable styles being Italian, Swedish, and Danish. Today, Wisconsin cheesemakers produce all three.

Fontinella   A cow milk cheese that should not be confused with Italian-made Fontina. Fontinella is distinctively sweet and creamy smooth, with just a hint of sharpness, and it melts beautifully.

Force Ripening   A method of speeding the ripening of a cheese by using a warmer environment than normal to naturally ripen the cheese.

Fresh   A term typically used to classify cheese varieties that have not been cured, such as mascarpone, cottage cheese, cream cheese, or ricotta. Cheeses that have been cured for very short periods, such as feta, may also be classified as fresh.

Fresh Mozzarella   A soft mozzarella with high-moisture content, meant to be eaten soon after it is produced.

Fromager   A French word to describe a person with in-depth knowledge of cheese. Sometimes spelled fromagier.

Fruity   A descriptive term for the sweet, fragrant aroma or flavor characteristic of certain semi-soft cheeses.

Gamey   A descriptive term for cheeses with strong flavors and penetrating aromas.

Gassy   A descriptive term for cheeses in packaging that becomes bloated. This may be a result of an increase in holding temperature or altitude, or it may indicate microbial production of carbon dioxide.

Goat   A classification of cheese made from goat milk.

Goaty   Distinctive flavor of cheeses made from goat milk.

Gorgonzola   An Italian-style blue cheese that is creamy-colored with greenish-blue veins. It can be semi-firm and crumbly or creamy and soft. Gorgonzola is typically produced in flatter wheels than the traditional Blue. See also Cheese of the Month, page 66.

Gouda   Gouda originated in Holland over 800 years ago. Made with whole milk and pale-yellow in color, Gouda has a rich, buttery, slightly sweet and nutty flavor and smooth, creamy texture, and develops complex caramel flavor and a firmer texture when aged. See also Cheese of the Month, page 28.

Grainy   (1) A descriptive term for gritty texture, which is desirable in certain hard-grating cheeses. (2) A flavor term that may be used to describe the grain-like (wheat) flavors that occur as the result of ripening.

Grana   The Italian term for hard-grating cheese such as Parmesan, Romano, Asiago, Parmigiano-Reggiano, Grana Padano, and Sapsago. See also American Grana®.

Grassy   A descriptive term for cheese taste that is related to the type of feed a cow has consumed prior to milking, such as silage, bitterweed, leeks, or onions.

Gruyère   Produced since the 11th century in the Alpine area between Switzerland and France, and made in Wisconsin and elsewhere today, Gruyère has a nutty, rich, full-bodied flavor, and firm texture.

Hard   A classification of cheese varieties exhibiting a relatively inelastic and unyielding texture, like Parmesan. Federal Standards of Identity state that firm cheeses have a maximum moisture content of 34 percent and a minimum milkfat content of 50 percent. Hard cheese such as Parmesan, Romano, and Asiago typically are well-aged and easily grated.

Havarti   A milder version of German Tilsit that was first made popular in Denmark. The Center for Dairy Research at the UW-Madison developed a special Wisconsin-style Havarti® that is firmer in texture and more buttery in flavor than other types and is available plain and flavored.

Imitation Pasteurized Process Cheese Spread   A cheese that possesses all the properties of pasteurized process cheese spread except the butterfat content is significantly lower than federal standards allow for labeling as a cheese spread.

Juustoleipa   Originally made from reindeer milk in northern Finland and Sweden. The name means "bread cheese." This "squeaky" cheese has a very buttery flavor, and the heat from baking caramelizes the sugars on the outside of the cheese to form a tasty crust similar to brown bread. Beneath the crust, the cheese is smooth and has a sweet, very mild flavor.

Kasseri   Originally from Greece and traditionally made of sheep milk and sometimes goat milk, Kasseri is also now made from cow milk. It has a mildly piquant, slightly tart flavor and a firm, slightly crumbly texture. Because of its ability to hold its shape when heated, Kasseri is often prepared grilled or fried, as in the classic Greek flaming dish, saganaki.

Lactic   (1) A general description applied to cheese exhibiting a clean, wholesome, milky, and slightly acidic flavor or aroma. (2) The type of organisms included in starter cultures for cheesemaking.

Lactose   A natural sugar found in most milks.

Lactose Intolerance   A human condition in which the digestive system is not able to properly break down the sugar lactose found in milk and dairy products. Common symptoms tend to be excessive gas and/or diarrhea.

Liederkranz   A cheese that originated in upstate New York in the late 1800s as a replica of a traditional soft, smelly cheese from Germany that immigrants missed and could no longer get because it would spoil during shipping. Related to Limburger, is

has the same texture and unique aroma, but features a distinctively robust and buttery flavor. Its moist, edible, golden-yellow crust cradles a pale-ivory interior with a honey-like consistency.

Limburger   First introduced in Belgium, Limburger later became associated with Germany, where this assertive cheese complemented the national taste for highly-flavored game and meats. Today, a single cheese plant in Monroe, Wisconsin, produces all the surface-ripened Limburger made in the United States. This famous "stinky" cheese has a pungent, strong aroma, a brownish surface and ivory interior, and an earthy, pungent flavor that increases in intensity with age.

Lipase   (1) An enzyme found in raw milk, also produced by microorganisms that split fat molecules into fatty acids. (2) Lipase flavor is a term also used to describe rancidity, especially where these flavors are desired in cheeses.

Mascarpone   A very soft, thick, smooth, and creamy-white cheese with a rich, buttery, slightly sweet flavor. Used most often as a dessert cheese, mascarpone nonetheless has many applications.

Mechanical Holes   Small, irregular openings in the body of cheese caused by manufacturing methods, not by gas fermentation. Colby, Brick, Muenster, and Monterey Jack are varieties with natural, mechanical openings. See also Open.

Medium-aged (Mellow)   Generally semi-firm, firm, or hard cheeses that have been cured for three to six months. Medium-aged cheeses are usually mellow and smooth textured. Frequently used to describe Cheddars.

Mild (Young)   A descriptive term for light, unpronounced flavors. Mild also refers to young, briefly-aged Cheddars.

Mold   (1) A condition created by the growth of various fungi during ripening, contributing to the individual character of cheese. Surface molds ripen from the rind inward. Internal molds, such as those used for blue-veined cheeses, ripen throughout the cheese. A moldy character can be clean and attractive, or unpleasantly musty or ammoniated. (2) A hoop or container in which cheese is shaped.

Monterey Jack   Scotsman David Jacks first produced Monterey Jack in Monterey, California in the 1890s. Available plain or flavored, it has a delicate, buttery, and slightly tart flavor and a creamy, open texture. It is an excellent melter.

Mouthfeel (Texture)   A general term for the "fabric" or "feel" of cheese when touched, tasted, or cut. Characteristics of cheese texture may be smooth, grainy, open or closed, creamy, flaky, dense, crumbly, and so forth, depending upon the specific variety.

Mozzarella   The classic pasta filata cheese, mozzarella is creamy white and smooth, has a mild, delicate, milky taste, and melts with a characteristic stretch. See also Pasta Filata and Cheese of the Month, page 126.

Muenster   Traditionally a washed-rind cheese, in the United States the rind may or may not be washed. Semi-soft, smooth, and elastic, Muenster usually has a bright-orange annatto coating and is mild when young and mellow with age.

Mushroomy   A descriptive term for ripened cheese, such as Brie, with an aroma and flavor similar to the clean, pleasant fragrance of mushrooms.

Natural   (1) A general classification for cheese that is made directly from milk. (2) The process whereby cheese is made directly from milk by coagulating or curdling the milk, stirring and heating the curd, draining the whey and collecting or pressing the curd.

Natural Rind   A rind that develops naturally on the cheese exterior through drying while ripening without the aid of ripening agents or washing.

Neufchatel   A lower-fat version of cream cheese. See also Cream Cheese.

Nutty   A descriptive term for cheese with a nut-like flavor, a characteristic of Swiss types.

Oaxaca   Named after the Oaxaca region of Mexico, where it originated, this is a Mexican-style pasta filata cheese, similar to string cheese or Provolone. See also Pasta Filata.

Off   A term referring to undesirable flavors or odors too faint or ill-defined to be more precisely characterized.

Oil Off   Refers to the separation of oil when cheese melts.

Oily   A descriptive term that may refer to body, aroma, and flavor. Cheese held out of refrigeration for extended periods may also appear oily.

Open   A term applied to cheese varieties containing holes that develop as a result of the manufacturing process.

Paneer   A fresh, farmer-style, and typically unsalted cheese originally from India. It is acid set and has vegetarian appeal because of the lack of animal rennet.

Paraffin   A wax coating applied to the rinds of some cheese varieties for both protection during export and extended life spans.

Part-skim   A term used to denote the manufacture of a cheese, such as mozzarella, with partly skimmed milk.

Parmesan   Known as the king of Italian cheeses, Parmesan has a granular texture, pale-yellow color, and a sweet, buttery, and nutty flavor that intensifies with age. See also Cheese of the Month, page 150.

Pasta Filata   Refers to a type of cheese wherein curds are heated and then stretched or kneaded before being molded into the desired shape. The resulting cheese has great elasticity and stretches when cooked or melted. Cheeses in this family include mozzarella, Provolone and string. See also Cheese of the Month, page 126.

Paste   A descriptive term for the interior texture of soft-ripened cheeses, such as Brie, that exhibit a semi-soft to runny consistency.

Pasteurization   The process of heating milk to a specific temperature for a specific period of time in order to destroy disease-producing bacteria.

Pasteurized Process Cheese   A blend of fresh and aged natural cheeses that have been shredded, mixed, and heated with an addition of an emulsifier salt, after which no further ripening occurs. Also called Processed or Process Cheese.

Pasteurized Process Cheese Food
A variation of pasteurized process cheese containing less fat and a higher moisture content. It differs from process cheese in that either nonfat dry milk or whey solids and water have been added, thus reducing the percentage of actual cheese in the finished product.

Pasteurized Process Cheese Spread
A variation of pasteurized process cheese containing a higher moisture content and lower milkfat content than process cheese food. A stabilizer is added to prevent separation of ingredients.

Pasture-grazed   A term describing cheese made exclusively from the seasonal milk of pasture-grazed animals.

Penicillium   Principal genus of fungi used to develop molds on certain cheese varieties during ripening.

Peppery   A descriptive term for cheese with a sharp, pepper flavor. .

Pressed Cheese   A descriptive term for cheese whose curd has been placed in a mold and literally pressed to form the intended shape of the finished cheese.

Process or Processed Cheese   See Pasteurized Process Cheese.

Provolone   A pasta filata cheese that is slightly piquant when young with a firm texture that becomes granular and more assertively flavored with age. Color is ivory to pale-beige. Available smoked. See also Pasta Filata and Cheese of the Month, page 126.

Pungent   A descriptive term for cheese with an especially poignant aroma or sharp, penetrating flavor.

Quark   This soft, fresh cheese has a mild flavor and a texture similar to ricotta.

Queso Blanco   A popular Mexican cheese that is fresh, crumbly, and slightly salty and will brown when heated but will not melt.

Queso Fresco   This popular Mexican-style cheese is snow-white, very soft, moist, and mild in flavor. It has a fine texture that makes it ideal for crumbling over salads and enchiladas, and stirring into refried beans and other cooked dishes. When heated, it softens and becomes creamy, but doesn't melt. Like other fresh cheeses, it is lower in fat and sodium (despite its salty flavor) than aged cheeses. It is sold in various shapes and sizes, most commonly in rounds, but it may also be packaged in cottage cheese-like tubs.

Queso Quesadilla   A rich, creamy melting cheese which originated in northern Mexico. This versatile cheese can be used in many traditional Mexican-style dishes and in place of any melting cheese. It is a part of a major group of Hispanic melting cheeses. See also Cheese of the Month, page 114.

Rancid   A term relating to flavors caused by the release of fatty acids from butterfat by the enzyme lipase. Some cheeses are not supposed to have flavors caused by fatty acids in high concentrations, such as Cheddar, while others, such as Romano, gain much of their flavor from the rancidity of fatty acids. See also Lipase.

Raw Milk   Milk that has not undergone pasteurization.

rBGH/Recombinant Bovine Growth Hormone   A synthetic version of bovine somatotropin (BST) used in dairy cows to increase milk production.

Rennet   An extract from the membranes of calves' stomachs that contains rennin, an enzyme that causes milk to coagulate and separate into curds and whey. Rennet-like enzymes, also used commercially, are produced by selected fungi and bacteria.

Ricotta   Italian cheesemakers originally produced ricotta (meaning recooked) from the whey that remained after making mozzarella and Provolone. They added lactic acid or vinegar to the whey and reheated it almost to boiling, causing the curds to precipitate and rise to the surface, where they were skimmed off and drained. Available in nonfat to whole milk variations, ricotta has a milky, delicate, mild, fresh flavor with just a hint of sweetness.

Ricotta Salata   A dry, salted ricotta cheese that has a sharp, almost tangy flavor. Use ricotta salata to dice into salads of all kinds. Its firm texture makes it perfect for tossing.

Rind   The outer surface of cheese. A rind varies in texture, thickness, and color. Cheeses may be rindless, display natural rinds, or possess rinds that are produced by harmless mold. See also Bloomy Rind and Natural Rind.

Rindless   Cheese without a rind. Some rindless varieties, such as Brick and Colby, are ripened (cured) in plastic film or other protective coating to prevent rind formation. Some cheeses, such as feta, are rindless because they are not allowed to ripen.

Ripe   A descriptive term for cheese that has arrived at peak flavor through aging.

Ripening   The chemical and physical alteration of cheese during the curing process. See also Aging and Curing.

Romano   Wisconsin cheesemakers make creamy-white, granular Romano with cow milk for a cheese that, like its Italian counterpart, has slightly more fat and a flavor that is sharp, tangy, and more assertive than Parmesan. See also Cheese of the Month, page 150.

Runny   A descriptive term for cheeses that have returned to a partially liquid state as a result of insufficient drainage of whey or exposure to excessive heat.

Salting   A step in the cheesemaking process requiring the addition of salt. Depending upon the cheese variety, salt can be added while the cheese is in curd form or rubbed on the cheese after it is pressed. Salt is used to help preserve cheese, as well as to enhance its flavor. Cheese also may be soaked in a salt solution, a process termed brining.

Salty   Most cheeses possess some degree of saltiness. Pronounced saltiness is characteristic of specific varieties; however, excessive saltiness is a defect. Cheeses lacking in salt are described as dull or flat.

Satiny   A descriptive term referring to the texture and mouthfeel of soft, spreadable cheese varieties. A satiny texture is characteristic of perfectly ripened Brie.

Semi-hard   A classification of cheese based upon body. Cheddar, Colby, Edam, and Gouda are examples of semi-hard cheese varieties.

Semi-soft   A wide variety of cheeses made with whole milk and which milt well when cooked. Cheeses in this category include Monterey Jack, Brick, Muenster, Fontina, and Havarti.

Sharp   A descriptive flavor term referring to the fully developed flavor of aged cheeses, such as Cheddar, Provolone, and some blue-veined varieties. The flavor is actually sharp and biting, but not excessively acrid or sour.

Sheep   A classification of cheese made from sheep milk.

Silky   See Satiny.

Smoked Cheese   Cheese that has been smoked in a process similar to smoking meat.

Soapy   Descriptive of a taste caused by long-chain fatty acids sometimes present in cheese caused by excessive milkfat breakdown. See Lipase and Rancid.

Soft-fresh   A category of cheeses with high moisture content that are typically direct set with the addition of lactic acid cultures. Cottage cheese and cream cheese are examples.

Soft-ripened   A classification of cheese based upon body. Brie and Camembert are examples.

Sour   A descriptive term for cheese with an excessive acid content. However, a mild, tangy sour flavor can be attractive in young cheeses.

Sour Milk Cheese   Cheese that has been curdled by natural souring or by the addition of lactic acid bacteria, such as cottage cheese. Sour milk cheese does not use rennet for coagulation.

Specialty Cheese   A subjective term used to classify cheeses of exceptional quality, notably unique or produced in quantities of less than 40 million pounds per year. Cheeses that are combinations of different cheese types also may be referred to as specialty.

Spiced   A term sometimes used to classify all cheese varieties containing spices, herbs, or flavorings. For example, caraway Gouda is a spiced cheese.

Spicy   A descriptive term for cheese varieties with a peppery, herby flavor.

Stabilizer   An ingredient added to a product to bind water, improve consistency, or stabilize an emulsion.

Starter   A culture that normally consists of varying percentages of bacterial or mold spores, or other microorganisms, lactic acid, enzymes, and natural chemicals. Starter cultures speed and control the process of curdling milk during cheesemaking in part by converting lactose to lactic acid. They also lend unique flavor characteristics to the cheese.

String Cheese   Typically made from mozzarella, string cheese is formed right after the cheese is made, when it is still soft and can be stretched. After being elongated it is brined and then cut into cylinders a few inches long and packaged. Nearly always consumed as a snack, string cheese is eaten by pulling apart the "fibers" of the cylinder.

Strong   A descriptive term for cheese with a pronounced or penetrating flavor and aroma.

Surface-ripened   A term referring to cheese that ripens from the exterior when a harmless mold, yeast, or bacteria is applied to the surface. Bloomy-rind cheeses, like Brie and Camembert, and washed-rind cheeses, like Limburger, are both surface-ripened.

Sweet Swiss   A rind cheese produced in Wisconsin that is a cross between Baby Swiss and Jarlsberg®, a Norwegian Swiss.

Swiss   This full-flavored, buttery, nutty cheese with characteristic holes is aged at least 60 days. See also Cheese of the Month, page 138.

Swiss-type   A term used to classify cheeses that share the common characteristics of eyes (holes) in their interior.

Thermalization   The process of heat-treating milk to less than 160 degrees F for less than 15 seconds prior to cheese production. This process utilizes a lower temperature for a shorter period of time than pasteurization.

Triple Cream   The French term for cheese which contains over 72 percent butterfat in the cheese solids. See also Creams and Fat Content.

Ultra Pasteurization   The process of super-heating milk to 275°F for 4 to 15 seconds to kill spores and extend shelf life.

Washed-rind   A cheese rind that has been washed periodically with brine, whey, beer, cider, wine, brandy, or oil during ripening. The rind is kept moist to encourage the growth of an orange-red bacteria. The bacteria may be scraped off, dried, or left to further rind development. Washed-rind and bloomy-rind cheeses compose what is termed the soft-ripening (surfaced-ripened) classification. Limburger is a washed-rind cheese.

Waxed   Prior to airtight shrink bags, cheesemakers would wrap their cheese in cheesecloth and dip in wax for preservation. Many wax colors denote some attribute of that cheese.

Weeping   A descriptive term referring to Swiss-type cheeses whose eyes glisten with bits of moisture. This is caused by the release of moisture by proteins as they are broken down during ripening. Weeping often indicates that a cheese has achieved peak ripeness and will exhibit full flavor. Can also be caused by storing cheese at too warm a temperature.

Whey   The thin, watery part of milk that separates from the coagulated curds during the first step of the cheesemaking process.

Selecting Cheese: Advice from the Pros

You're standing before a cheese counter that's chock-a-block with Wisconsin varieties. What should you know? How do you choose? How can you best use your selection? Here are words of wisdom from cheese experts around the state:

VICKI THINGVOLD, MEISTER CHEESE COMPANY, MUSCODA
- Be adventurous; try something out of your comfort zone.
- Pair it with what is in season.

HEIDI BUHLMAN-GRUNDVIG, EAU GALLE CHEESE FACTORY, DURAND
- If possible, sample before you buy.
- Buy smaller quantities more often.

DEBBIE CRAVE, CRAVE BROTHERS FARMSTEAD CHEESE, WATERLOO
- Try new applications; example: mascarpone pizza with figs, candied bacon, and sweet onions.

JOSEPH WIDMER, WIDMER'S CHEESE CELLARS, THERESA
- Buy it fresh, taste before you buy, and look for artisan cheeses.

JOE BURNS, BRUNKOW CHEESE OF WISCONSIN, DARLINGTON
- Make sure you don't get stuck with a disproportion of rind on the cheese.
- In most cases, white wine and/or beer pairs better with cheese than red wine.

KEN HEIMAN, NASONVILLE DAIRY, MARSHFIELD
- Start with a mild flavor and work your way to more distinct flavors.

KEVIN, PATRICK, AND BRIAN MCCLUSKEY, MCCLUSKEY BROTHERS AT SHILLELAGH GLEN FARMS, HILLPOINT
- Know your farmer and his/her farming practices. This really matters.
- Know your milk source. Pay attention to what the animals are eating and when. A 100 percent grass-fed cheese made from a herd on spring pasture, for example, will have a different flavor than that same cheese sourced from the same herd on fall pasture, and both of these cheeses will be different from that same cheese sourced from a herd fed grain.
- Know your cheesemaker.

ADAPTED FROM "CHEESE 101" AT EATWISCONSINCHEESE.COM
- Cheese should have a fresh, clean appearance with no cracks or surface mold.
- Be sure the packaging is sealed properly, without any openings or tears.
- Buy cheese at a store where frequent shipments are delivered. Check the "use by" or "sell by" dates on packaged cheese. If buying fresh-cut cheese, ask the clerk how best to wrap it for storage as well as how long the cheese can be kept.

Handling and Storing Cheese: Advice from the Pros

Paper or plastic? To freeze or not to freeze? What are the best ways to care for cheese? Here's counsel to help you keep your purchases in peak condition longer:

ADAPTED FROM "CHEESE 101" AT EATWISCONSINCHEESE.COM

- Cheese loses flavor and moisture when exposed to air. Wrap hard cheeses such as Parmesan in tightly drawn plastic wrap. Soft or fresh cheeses, such as mascarpone, are best stored in clean, airtight containers. Semi-hard cheeses, including Cheddar and Gouda, can be wrapped in plastic wrap or parchment.

- Refrigerate cheese between 34° and 38°F. Because it easily absorbs other flavors, keep cheese away from other aromatic foods in the refrigerator.

- Most cheese is easiest to cut when chilled. However, some hard cheeses cut better when they are brought to room temperature.

- Before eating or serving cheese, trim off any dry edges or surface mold.

- Natural and pasteurized process cheese should last about four to eight weeks in the refrigerator, while fresh and grated hard cheese with higher moisture content should be used within two weeks.

- When freezing cheese, wrap pieces tightly in weights of one pound or less. Thaw cheese in the refrigerator and use it within a couple of days.

- A cheese that has been frozen is best used as an ingredient. The best candidates for freezing are firm cheeses, such as Swiss, and hard cheeses, such as Parmesan. Semi-soft and hard cheeses will be more crumbly after freezing, while softer cheeses will separate slightly.

KERRY HENNING, HENNING CHEESE, KIEL

- When buying a vacuum package of blue cheese, open the package and let air contact the cheese. The natural molds tend to go dormant when vacuumed, but return to their natural state when exposed to air.

- If you feel your cheeses mold faster than they should, look at how clean your refrigerator is. That moldy food in the back will easily contaminate everything else. Use straight vinegar to wipe down all surfaces of the fridge. Vinegar destroys mold spores and leaves the fridge smelling fresh.

TONY HOOK, HOOK'S CHEESE COMPANY, MINERAL POINT

- Freezing will change the texture of cheese. Most cheeses should not be frozen. There are a few that can be. You can freeze blue cheeses if you are going to crumble them. You can also freeze Parmesan if you are going to grate it.

Map of Cheese Factories and Cheese Stores

Enjoy a taste of Wisconsin! Use this map to find cheese plants and cheese shops in the northwest quadrant of Wisconsin. An orange cheese wheel icon indicates the site of a cheese-producing plant with a retail store. A green cheese wedge indicates the site of a cheese shop, or a retail food store that specially features cheese.

Look for the numbers on the map that are in the area you're traveling to or interested in, then refer to the corresponding numbers in the list that follows this page.

This map provides a general reference for location and is not meant to be used for navigation. Please contact the businesses or check their websites for specific directions, as well as hours of business and other important information.

Map and listing are excerpted and adapted from eatwisconsincheese.com, the website of the Wisconsin Milk Marketing Board.

1 Associated Milk Producers, Inc. (AMPI)

14193 Cty. Rd. S, Jim Falls
715.382.5169
Unique selection of cheeses including several pepper-style cheeses such as the award-winning Pepper Jack made in the Jim Falls plant. Retail Store.

2 Bass Lake Cheese Factory

598 Valley View Tr., Somerset
715.247.5586
blcheese.com
Handmade, specialty cheeses including Butter Jack, Juustoleipa, Merlot Cheddar, and award-winning Muenster Del Ray. Retail Store. Observation Window. Mail Order. Online Orders.

3 Benoit Cheese

23920 Cty. Hwy. F, Ashland
715.746.2561
benoitcheese.com
Over 150 varieties of Wisconsin cheese. Free samples every day all day. Mail Order.

4 Bolen Vale Cheese

E977 State Rd. 64, Downing
715.265.4409
bolenvalecheese.com
Full line of dairy products; bulk spices and baking supplies; local, natural meats; pizza; craft items; and more. Retail Store. Mail Order. Online Orders. Tours by appointment.

5 Burnett Dairy Cooperative

11631 State Rd. 70, Grantsburg
715.689.2468
burnettdairy.com
Wood River Creamery line of handcrafted artisan cheese, as well as award-winning string cheese, Provolone, mozzarella, Colby, and more. Ice cream counter and bistro. Retail Store. Observation Window. Mail Order. Online Orders.

6 Cady Cheese Factory, Inc.

126 Hwy 128, Wilson
715.772.4218
cadycheese.com
Large cheese selection including award-winning Longhorn Colby, Cheddar, Monterey Jack, Gold'n Jack, Hot Pepper, and more, plus sausage and gifts. Retail Store. Observation Window. Mail Order. Online Orders. Tours.

7 Comstock Creamery

1858 Hwy. 63, Comstock
715.822.2437
ellsworthcheesecurds.com
Home to Blaser's award-winning cheeses. Known for Muenster, Havarti, Monterey Jack, Cheddar, and rbST-free Valley View of Ellsworth White Cheddar. Also sells a wide variety of Wisconsin cheese, wine, ice cream, fudge, and gifts, and operates a deli. Retail Store. Observation Window. Mail Order. Online Orders.

8 Eau Galle Cheese Factory

N6765 State Hwy. 25, Durand
715.283.4211
eaugallecheese.com
Large variety of cheese including award-winning Parmesan plus Wisconsin wines, maple syrup, jams, mustards, and sausages. Retail Store. Mail Order. Online Orders.

9 Ellsworth Cooperative Creamery

232 N. Wallace St., Ellsworth
715.273.4311
ellsworthcheesecurds.com
Over 240 varieties of Wisconsin cheese, along with snack meats, Cheddar curds daily at 11 a.m, and product sampling. Retail Store. Mail Order. Online Orders.

10 Falcon Foods – UW River Falls

410 S. 3rd St., River Falls
715.425.3704
www.uwrf.edu/ANFS/FalconFoods/Index.cfm
Dairy products, meats, student-crafted cheese and ice cream made with milk from the University Dairy Farm. Retail Store. Observation Window. Mail Order.

11 Foster Cheese Haus

E10934 Cty. Hwy. HH, Osseo
715.597.6605
fostercheesehaus.com
Over 100 Wisconsin cheeses: fresh cheese curds, organic, aged Cheddar, Bleu, Gouda, Goat, Sheep and more. Mail Order. Online Orders. Restaurant.

12 Gad Cheese, Inc.
2401 Cty. Hwy. C, Medford
715.748.4273
Forty varieties of cheese, fresh cheese curds, and deep-fried cheese curds. Mail Order.

13 Gingerbread Jersey
1025 W. Lincoln St., Augusta
715.286.4007
gingerbreadjersey.com
Award-winning cheeses including Gouda, Cheddar, Colby, Horseradish Cheddar, smoked mozzarella, and cheese curds. Retail Store. Tours.

14 Hawkeye Dairy Store
118 S. Fourth St., Abbotsford
715.223.6358
hawkeyedairy.com
Over 100 varieties of Wisconsin cheese and cheese curds plus 28 flavors of ice cream, sausage, maple syrup, and other products. Mail Order. Online Orders.

15 Holland's Family Cheese
N13851 Gorman Ave., Thorp
715.669.5230
hollandsfamilycheese.com
Authentic, artisan Gouda cheeses made from farmstead milk and ingredients imported from Holland. Also have traditional Dutch items and delicacies. Retail Store. Observation Window. Mail Order. Online Orders. Tours available upon request.

16 LaGranders Hillside Dairy
W11299 Broek Rd., Stanley
715.644.2275
lagranderscheese.com
Award-winning Colby and Monterey Jack overseen by a Master Cheesemaker plus Colby Longhorns, Monterey Jack, Co-Jack, Pepper Jack, and fresh cheese curds. Retail Store. Mail Order.

17 Lynn Dairy, Inc.
W1929 US Hwy. 10, Granton
715.238.7129;
lynndairy.com
Fourth-generation factory with international award-winning Monterey Jack plus Cheddars, feta, Marble,

Pepper Jack, and more. Retail Store. Observation Window. Mail Order. Online Orders.

18 Market on Sixth
113 W. 6th St., Marshfield
715.387.2000
marketonsixth.com
Wisconsin farmstead and artisan cheeses cut to order.

19 Miller's Cheese House
2248 Hammond Ave., Rice Lake
715.234.4144/800.677.4144
millerscheesehouse.com
Over 70 varieties of Wisconsin cheese plus homemade fudge. Mail Order. Online Orders.

20 Nasonville Dairy, Inc.
10898 Hwy. 10 W., Marshfield
715.676.2177
nasonvilledairy.com
Oldest cheese plant in Wood County, featuring award-winning feta, Colby, Marble, Jack. Retail Store. Observation Window. Mail Order. Online Orders. Tours.

21 Nasonville North
N14505 Sandhill Ave., Curtiss
715.223.3338
nasonvilledairy.com
Best known for cheese curds. Retail Store. Observation Window. Mail Order.

22 Nelson Cheese Factory
S237 St. Hwy. 35, Nelson
715.673.4725
nelsoncheese.com
Historic 1800s factory building in scenic Mississippi River Valley. Cheese, fine wines, gourmet food, ice cream cones, deli, and patio. Mail Order. Online Orders.

23 North Hendren Co-op Dairy
W8204 Spencer Rd., Willard
715.267.6617
northhendrenbluecheese.com
Winner of many awards for Black River Blue and Gorgonzola. Retail Store.

 **Pavilion Cheese and Gifts**
1201 E. Division St., Neillsville
715.743.3333
cwbradio.com/cheesegifts
Assortment of specialty cheeses crafted in Wisconsin and wine offerings.

 **Weber's Farm Store**
9706 Cty. Rd. H, Marshfield
715.384.5639
Farmstead milk, cheese, eggs, butter, ice cream, meat, heavy cream, and more. Onsite milk processing. Observation Window.

 **Welcome Dairy, Inc.**
H4489 Maple Rd., Colby
715.223.2874
welcomedairy.com
Natural and process cheese; best known for Hot Pepper Cheese, Colby, and aged Cheddar. Retail Store.

 **Yellowstone Cheese, Inc.**
24105 Cty. Hwy. MM, Cadott
715.289.3800
yellowstonecheese.com
Featuring Colby, Monterey Jack, Cheddar, and curds made from Grade A, BGH-free milk from a single farm. Retail Store. Observation Window. Mail Order. Online Orders. Tours.

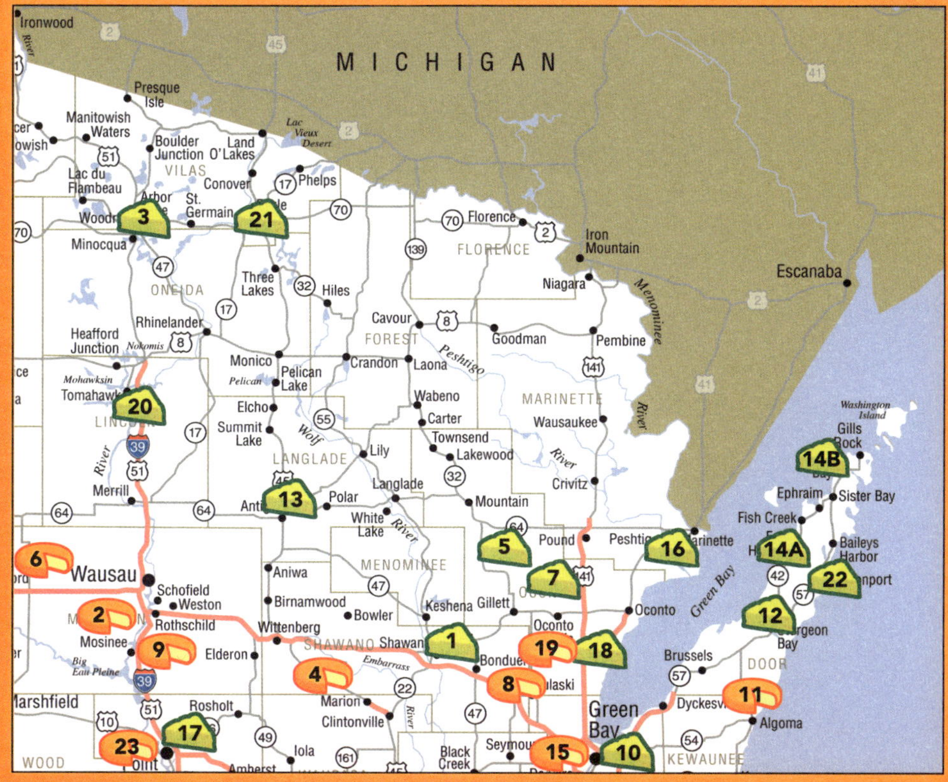

Map of Cheese Factories and Cheese Stores

Enjoy a taste of Wisconsin! Use this map to find cheese plants and cheese shops in the northeast quadrant of Wisconsin. An orange cheese wheel icon indicates the site of a cheese-producing plant with a retail store. A green cheese wedge indicates the site of a cheese shop, or a retail food store that specially features cheese.

Look for the numbers on the map that are in the area you're traveling to or interested in, then refer to the corresponding numbers in the list that follows this page.

This map provides a general reference for location and is not meant to be used for navigation. Please contact the businesses or check their websites for specific directions, as well as hours of business and other important information.

This map is excerpted and adapted from eatwisconsincheese.com, the website of the Wisconsin Milk Marketing Board.

1 **Belle Plaine Cheese**
N3473 Wisconsin Ave., Shawano
715.526.2789/866.245.5924
Seventy-five flavors of Wisconsin cheese shipped anywhere in the U.S. from October to May. Custom-made gift boxes. Mail Order.

2 **Bletsoe Cheese, Inc.**
8281 3rd Ln., Marathon
715.443.2526
Fresh squeaky curds and other cheeses. Retail Store. Mail Order.

3 **Cheese Board of Wisconsin**
8524 Hwy. 51 N., Minocqua
877.230.1338
thecheeseboard.com
Large cheese selection, as well as meats, candies, snacks, sauces, mustards, and other specialty foods. Mail Order. Online Orders.

4 **Dupont Cheese, Inc.**
N10140 Hwy. 110, Marion
715.754.5424
dupontcheeseinc.com
Large variety of cheese, cheese curds, and other Wisconsin products. Retail Store, Observation Window. Mail Order. Online Orders.

5 **G.G.'s Cheese Mart LLC**
10907 Hwy. 32, Suring
920.842.3219
ggscheesemart.com
Wide variety of Wisconsin cheese, Hotsticks, sausage, Hansen's ice cream, smoked salmon and trout, and more. Cheese trays, gift boxes, and baskets made to order. Mail Order.

6 **Harmony Specialty Dairy Foods**
R754 Spruce Ln., Athens
715.687.4236
harmonyho.com
Wide assortment of cheese including a line of British-style cheeses under the Golden Age Cheese label (Gloucester, Caerphilly, Cheshire, Abergele) and a line of Stark Kosher. Beer and wine tastings. Retail Store. Mail Order. Online Orders. Tours available by appointment.

7 **Kugel's Cheese Mart**
311 North Rosera St., Lena
920.829.5537
kugelscheese.com
Large selection of factory-direct cheese with daily specials and sampling available. Mail Order. Online Orders.

8 **Laney Cheese**
W1031 Main Laney Dr., Pulaski
920.822.5432
laneycheesestore.com
Third-generation family factory featuring aged Cheddars, string cheese whips, Colby, and fresh curds daily. Retail Store. Mail Order. Online Orders.

9 **Mullins Cheese, Inc.**
598 Seagull Dr., Mosinee
715.693.3205
mullinscheese.com; mullinswhey.com
Dozens of cheese varieties, Whey Protein Isolate Powder, ice cream, and gifts. Retail Store. Video tours.

10 **Nala's Fromagerie**
2633 Development Dr., Suite 30, Green Bay
920.347.0334
nalascheese.com
European-style cheese store offering over 160 varieties, plus olives and cured meats. Mail Order. Online Orders.

11 **Renard's Cheese**
248 Cty. S., Algoma
920.487.2825
renards.com
Specialties include aged Cheddar, Colby, string cheese, and fresh cheese curds daily. Retail Store. Mail Order. Online Orders.

12 **Renard's Cheese**
2189 Cty. DK, Sturgeon Bay
920.825.7272
renardscheese.com
Wide variety of cheese, cheese curds, homemade sausage, and specialty foods from Door County. Mail Order. Online Orders.

 ### Sartori Co.
201 Morse St., Antigo
715.623.2301
sartoricheese.com
Over 100 national and international awards for artisan cheese, including original BellaVitano. Retail Store. Online Orders.

 ### Schoolhouse Artisan Cheese
schoolhouseartisancheese.com
Over 30 Wisconsin artisan cheeses plus dry-cured meats. Mail Order. Online Orders.
Two locations:
7813 Hwy. 42, **14A Egg Harbor**
920.868.2400
12042 Hwy. 42, **14B Ellison Bay**
920.854.6600

 ### Scray Cheese Co., LLC
2082 Old Martin Rd., DePere
920.347.0303
scraycheese.com
Four generations of award-winning Smoked Gouda and Cheddar, plus a wide variety of cheese & sausage, and more. Retail Store. Observation Window. Mail Order. Online Orders.

 ### Seguin's House of Cheese
W1968 US Hwy. 41, Marinette
800.338.7919
seguinscheese.com
Wisconsin cheese, local preserves, wines, and extra-hot mustard. Retail Store. Mail Order. Online Orders.

 ### Ski's Meat Market
5370 Hwy. 10 E., Stevens Point
715.344.8484
skismeatmarket.com
Upscale meats and more than 100 varieties of Wisconsin artisan and farmstead cheese.

 ### Springside Cheese and Wine
5068 Hwy. 141, Suite 1105, Oconto
920.834.6395
springsidecheese.com
Large selection of specialty cheeses, gift boxes, wines, and Wisconsin products. Online Orders.

 ### Springside Cheese Factory and Store
7989 Arndt Rd., Oconto Falls
920.829.6395
springsidecheese.com
Family-owned, award-winning specialty cheese factory producing hand-made traditional Cheddar, Colby, and Monterey Jack. Retail Store. Observation Window. Mail Order. Online Orders.

 ### The Cheese Shoppe
112 W. Wisconsin Ave., Tomahawk
715.224.2627
thecheeseshoppeonline.com
Traditional and specialty Wisconsin cheese, gourmet foods, and wines. Online Orders.

 ### The Flour Sack
348 Pine St., Eagle River
715.479.7249
floursack.com
Artisan Wisconsin cheeses, jams, mustards, salsa, and snacks. Mail Order. Online Orders. Café.

 ### Wisconsin Cheese Masters
4692 Rainbow Ridge Ct., Egg Harbor
920.868.4320
wisconsincheesemasters.com
Door County's premier stop for Wisconsin cheese, with sampling and private tastings. Mail Order. Online Orders. Tours.

 ### Wisconsin Dairy State Cheese Co.
Hwy. 13/34, Rudolph
715.435.3144
Making bulk Cheddar, Colby, curds, and Monterey Jack, and selling over 200 varieties of Wisconsin cheese. Retail Store. Observation Window. Mail Order. Tours by appointment.

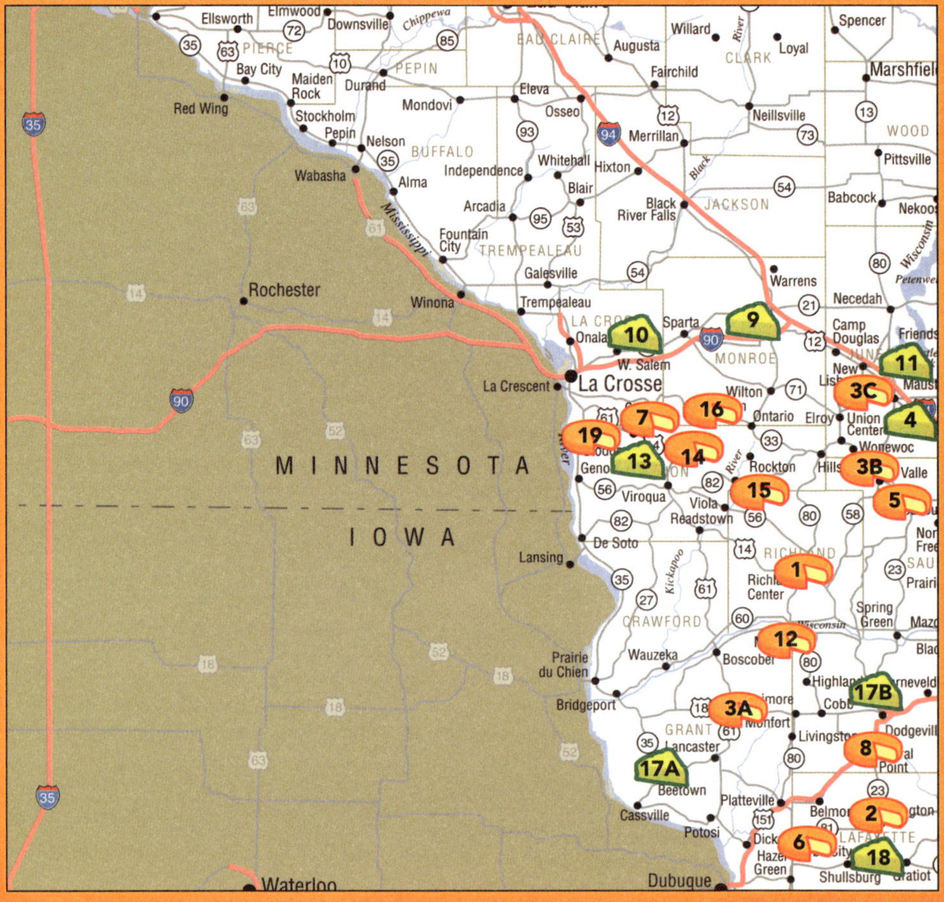

Map of Cheese Factories and Cheese Stores

Enjoy a taste of Wisconsin! Use this map to find cheese plants and cheese shops in the southwest quadrant of Wisconsin. An orange cheese wheel icon indicates the site of a cheese-producing plant with a retail store. A green cheese wedge indicates the site of a cheese shop, or a retail food store that specially features cheese.

Look for the numbers on the map that are in the area you're traveling to or interested in, then refer to the corresponding numbers in the list that follows this page.

This map provides a general reference for location and is not meant to be used for navigation. Please contact the businesses or check their websites for specific directions, as well as hours of business and other important information.

This map is excerpted and adapted from eatwisconsincheese.com, the website of the Wisconsin Milk Marketing Board.

1 **Alcam Creamery Co., Inc.**
300 S. Main St. (Hwy. 80),
  Richland Center
608.647.4177
alcamcreamery.com
*Butter in many sizes and shapes;
specialty is hand-rolled butter. The
store is down the street from the
factory and includes over 30 cheese
varieties. Retail Store. Mail Order.
Online Orders.*

2 **Brunkow Cheese of Wisconsin**
17975 Cty. F, Darlington
608.776.3716
brunkowcheese.com
*Cheese curds, cold pack cheese spread,
and Brun-uusto™ assorted cheeses.
Retail Store. Mail Order.*

Carr Valley Cheese Co., Inc.
carrvalleycheese.com
*Eight retail locations, three associated
with factories, offering over 90 award-
winning artisan cheeses:*

3A 675 Lincoln Ave., **Fennimore**
888.499.3778
*Retail Store. Mail Order.
Online Orders.*

3B S3797 County Rd. G, **LaValle**
608.986.2781
*Watch award-winning Cheddar
and artisan cheeses being made!
Retail Store, Observation Window.
Mail Order. Online Orders. Tours.*

3C 1042 E. State St., **Mauston**
608.847.6632
*Retail Store. Observation Window.
Mail Order. Online Orders.*

4 1002 State Highway 82, **Mauston**
608.847.4891
Mail Order. Online Orders.

5 **Cedar Grove Cheese**
E5904 Mill Rd., Plain
800.200.6020
cedargrovecheese.com
*Traditional and artisan cheeses, fresh
cheese curds. Observation Window.
Mail Order. Tours.*

6 **Gile Cheese, LLC/Carr Cheese Factory**
116 N. Main St., Cuba City
608.744.3456
gilecheese.com
Champion cheeses including Colby,
Baby Swiss, Cheddar, Monterey Jack
plus many other flavored cheeses
and Cheddar curds. Retail Store. Mail
Order. Online Orders.

7 **Hidden Springs Creamery**
S1597 Hanson Rd., Westby
608.634.2521
hiddenspringscreamery.com
Award-winning fresh handmade
sheep milk and mixed milk aged
cheeses. Online Orders. Observation
Window. Retail and tours available by
appointment.

8 **Hook's Cheese Co., Inc.**
320 Commerce St., Mineral Point
608.987.3259
hookscheese.com
*Four styles of World Champion Blue
Cheese cured in on-site underground
cave, artisanal Cheddars aged to 10 and
12 years, Tilston Point, Colby, Monterey
Jack, Swiss, Parmesan and Sweet
Constantine. Retail Store. Mail Order.*

9 **Humbird Cheese Mart**
2010 Eaton Ave., Tomah
608.372.6069/888.684.5353
humbirdcheese.com
*Aged Cheddars as well as over 100
other varieties of Wisconsin cheese,
homemade fudge, ice cream, and
souvenirs. Retail Store. Mail Order.
Online Orders.*

10 **Le Coulee Cheese Castle**
112 S. Leonard St., West Salem
608.786.2811
lecouleecheese.com
*Over 50 varieties of Wisconsin cheese,
hand-dipped ice cream cones, and
Wisconsin products and souvenirs.
Mail Order.*

 **Log Cabin Deli & Cheese Store**
517 State Rd. 82E, Mauston
608.548.4610
logcabindeliandcheese.com
Over 60 varieties of Wisconsin cheese, specialty foods, and made-to-order sandwiches. Mail Order. Online Orders.

 **Meister Cheese Co.**
1050 Industrial Dr., Muscoda
608.739.3134
meistercheese.com
Monterey Jacks and Cheddars made with high-quality milk from local farms using sustainable practices. Retail Store. Observation Window. Mail Order. Online Orders.

 **Nordic Creamery**
S2244 Langaard Lane, Westby
608.634.3199
nordiccreamery.com
wisconsinbutter.com
Artisan cheese from goat and cow milks and the only specialty butter plant in the Midwest. Retail Store. Observation Window. Mail Order. Online Orders. Tours.

 **Old Country Cheese**
5510 Cty. Hwy. D, Cashton
888.320.9469
oldcountrycheese.com
Cheese made from Amish farm milk, including curds, Juustoleipa, Cheddar, Marble, Colby, and more. Retail Store. Observation Window. Mail Order. Online Orders.

 **Organic Valley/CROPP Cooperative**
La Farge
608.625.2602
organicvalley.coop
North America's largest farmer-owned organic cooperative and brand, producing award-winning organic cheeses, milk, and other dairy products. Retail store located at 507 W. Main St., La Farge. Tours.

 **Pasture Pride Cheese, LLC**
110 Eagle Dr., Cashton
608.654.7444
pasturepridecheese.com
Amish milk cheese, including award-winning Juusto Baked Cheese, and Amish crafts. Retail Store. Observation Window. Mail Order. Online Orders. Tours.

Schurman's Wisconsin Cheese Country, Inc.
schurmanscheese.com

 7786 Cty. U E., **Beetown**
608.794.2422
Aged Cheddar, fresh curds, wine, and more. Mail Order. Online Orders.

 1401 Hwy. 23 N., **Dodgeville**
608.935.5741
Many local specialties including Wisconsin cheese and beer. Fresh-from-the-vat curds on Mondays and Fridays. Mail Order. Online Orders.

 **Shullsburg Creamery Cheese Store**
208 W. Water St., Shullsburg
608.965.3855/800.533.9594
shullsburgcheesestore.com
Quaint cheese store in an historic district. Also meats and wines. Mail Order. Online Orders.

 **Westby Cooperative Creamery**
401 S. Main St., Westby
608.634.3183
westbycreamery.com
Certified rBST-free creamery producing cottage cheese, butter, hard cheese, yogurt, sour cream, and more. Retail Store. Observation Window. Mail Order. Online Orders.

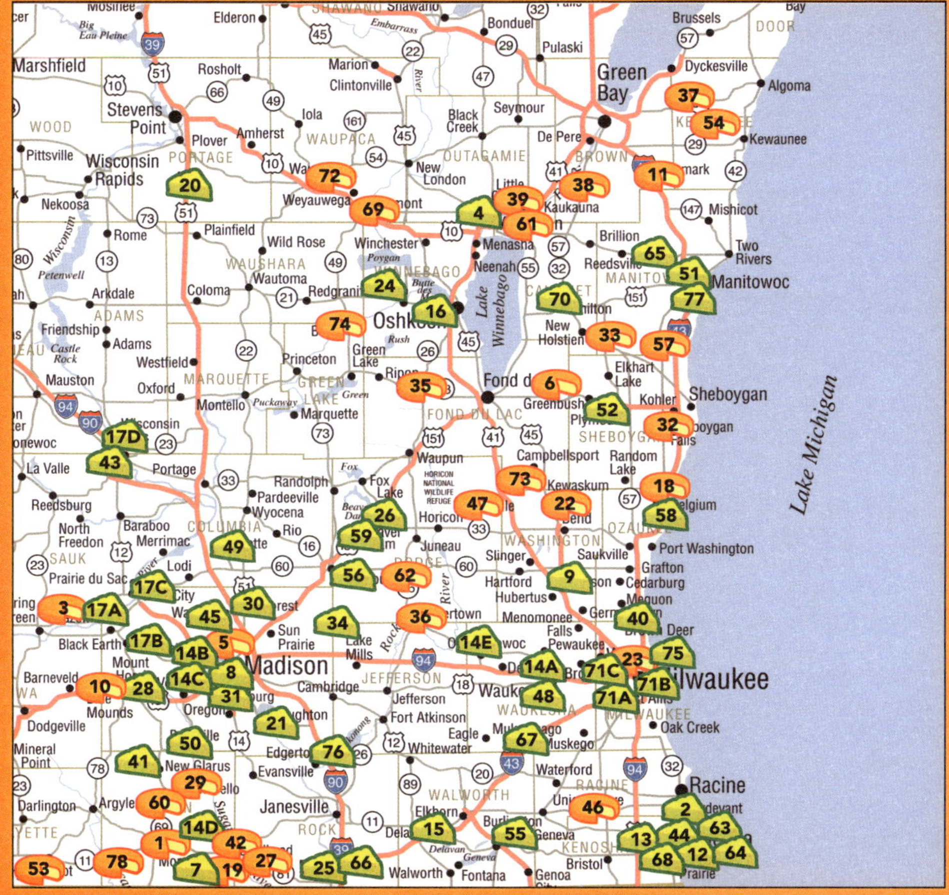

Map of Cheese Factories and Cheese Stores

Enjoy a taste of Wisconsin! Use this map to find cheese plants and cheese shops in the southeast quadrant of Wisconsin. An orange cheese wheel icon indicates the site of a cheese-producing plant with a retail store. A green cheese wedge indicates the site of a cheese shop, or a retail food store that specially features cheese.

Look for the numbers on the map that are in the area you're traveling to or interested in, then refer to the corresponding numbers in the list that follows this page.

This map provides a general reference for location and is not meant to be used for navigation. Please contact the businesses or check their websites for specific directions, as well as hours of business and other important information.

This map is excerpted and adapted from eatwisconsincheese.com, the website of the Wisconsin Milk Marketing Board.

1 **Alp and Dell Cheese Store**
657 2nd St., Monroe
608.328.3355
alpanddellcheese.com
Associated with Emmi-Roth Kase cheese factory. Cheese, meat, ice cream, candy, and cheese-related gifts. Retail Store. Mail Order. Online Orders. Observation Window. Tours.

2 **Apple Holler**
5006 S. Sylvania Ave., Sturtevant
800.238.3629
appleholler.com
Cheese, cheese boxes, caramel apples, apple pies, apple dumplings, apple everything! Mail Order. Online Orders.

3 **Arena Cheese**
300 Highway 14, Arena
608.753.2501
arenacheese.com
Home of the original Co-Jack cheese; known for Colby Jack, Colby, and fresh cheese curds. Retail Store. Observation Window. Mail Order. Tours.

4 **Arthur Bay Cheese Co.**
237 E. Calumet St., Appleton
920.733.1556
Fresh cheese curds, a variety of traditional and artisanal Wisconsin cheeses, including some award-winning cheeses, many other quality Wisconsin products.

5 **Babcock Hall Dairy Store**
1605 Linden Dr., Madison
608.262.3045
babcockhalldairystore.wisc.edu
Large variety of cheese and flavors of ice cream. Also serve breakfast and lunch. Retail Store. Observation Window. Online Orders. Tours (608.265.9500).

6 **Baker Cheese, Inc.**
N5279 County Rd. G, St. Cloud
920.477.7111
bakercheese.com
Award-winning fresh string cheese sold daily and many other fine Wisconsin cheeses. Retail Store. Mail Order.

7 **Baumgartner's Cheese Store and Tavern**
1023 16th Ave., Monroe
608.325.6157
baumgartnercheese.com
Wisconsin's longest running cheese store. Great cheese, great beer, great food. Mail Order. Online Orders.

8 **Bavaria Sausage and Cheese Chalet**
6317 Nesbitt Rd., Madison
608.271.1295
bavariasausage.com
Award-winning authentic German sausage; over 100 types of sausage, brats, hams, etc., Wisconsin and imported cheeses, including Cheddar aged 2-17 years. Mail Order. Online Orders.

9 **Bieri's Jackson Cheese and Deli**
3271 Highway P, Jackson
262.677.3227
bierischeese.com
Over 150 varieties of cheese including 7- and 11-year Aged Cheddar, Aged Brick, and Hickory Nut Muenster. Mail Order.

10 **Bleu Mont Dairy**
3480 Cty. Hwy. F, Blue Mounds
608.767.2875
cheeseforager.com/bleumont
Organic pasteurized and raw-milk artisan cheeses from grass-fed cows. Award-winning Bandaged Cheddar, surface-cured in an underground aging cellar. Retail at Dane County Farmers' Market. Tours by appointment.

11 **Blossoms and Steve's Cheese Co.**
220 S. Bohemia Dr., Denmark
920.863.2397/888.427.0483
stevescheese.com
Many varieties of Wisconsin cheese, including a full line of BelGioioso cheeses. Also fresh curds, wines, fudge, and preserves. Mail Order. Online Orders.

12 **Bobby Nelson's Cheese Shop**
2924-120th Ave., Kenosha
262.859.2232
Cheese and other Wisconsin specialty products. Mail Order.

 **Borzynski's Farm and Floral Market**
11600 Washington Ave., Mount Pleasant
262.886.2235
borzynskis.com
Wisconsin cheese and many other products.

Brennan's Farm Market
brennansmarket.com
Five retail locations specializing in Wisconsin cheese, premium fruit, vegetables, micro beer and new-world wine. Online Orders.

 19000 W. Bluemound Rd., **Brookfield**
262.785.6606

 5533 University Ave, **Madison**
608.233.2777

 8210 Watts Rd., **Madison**
608.833.2893

 701 8th St., Monroe, **Monroe**
608.325.4433

 1670 Old School House Rd., **Oconomowoc**
262.569.0067

 **Brick Street Market**
104 E. Walworth Ave., #101, Delavan
262.740.1880
brickstreetmarket.com
Unique collection of cheese from Wisconsin and beyond. Mail Order. Online Orders.

 **Butcher Block Meats and Cheese**
234 N. Koeller St., Oshkosh
920.230.2220
bbmcoshkosh.com
Butcher shop also specializing in artisan Wisconsin cheese. Featuring cut-to-order and pre-wrapped cheeses, honey, preserves, crackers, and other Wisconsin specialty foods. Retail Store.

Carr Valley Cheese Co., Inc.
carrvalleycheese.com
Eight retail locations, three associated with factories, offering over 90 award-winning artisan cheeses:

 428 Wall St. **Mazonmanie**
608.401.1043
Mail Order. Online Orders.

 2831 Parmenter St., Suite 160, **Middleton**
608.824.2277
Mail Order. Online Orders.

 807 Phillips Blvd., **Sauk City**
608.643.3441
*Home of the Carr Valley Cooking School.
Mail Order. Online Orders.*

 420 Broadway St., **Wisconsin Dells**
608.254.7200
Mail Order. Online Orders.

 **Cedar Valley Cheese**
W3115 Jay Rd., Belgium
920.994.9500/877.i8curds
cedarvalleycheesestore.com
Specializing in mozzarella, Provolone, and award-winning, hand-pulled string cheese. Also carry Wisconsin products and souvenirs, wines, ice cream, deep fried mozzarella sticks, and curds. Retail Store. Observation Window. Mail Order. Online Orders.

 **Chalet Cheese Cooperative**
N4858 Hwy. N Monroe
608.325.4343
Only plant in U.S. making Limburger; also featuring` award-winning Swiss and Baby Swiss. Retail Store.

 **Cheese Pleasers Inc**
5364 Glennville Rd., Bancroft
715.335.6750
cheesepleasersinc.com
Large variety of cheese, gift boxes, snacks and cheese curds. Mail Order.

Cheesers

183 E. Main St., Stoughton
608.873.1777
cheesers.com
Over 130 varieties of cheese; fresh curds, lefse, and Babcock Hall ice cream. Mail Order.

Cheeseville Cheeseboxes

7660 Trading Post Trail, West Bend
262.692.9443/262.483.4425
cheesevillecheeseboxes.com
Holiday cheese boxes available year round. Mail Order. Online Orders. Tours.

Clock Shadow Creamery

138 W. Bruce St., Milwaukee
414.273.9711
clockshadowcreamery.com
Milwaukee's first and currently only cheese factory. Fresh cheese, Cheddar curds, gift boxes, and artisan ice cream. Retail Store. Observation Window. Mail Order. Tours.

Colonial Cheese House and Gifts, Inc.

230 W. Main St., Omro
920.685.6570/800.985.6590
omrocheesehouse.com
Eighty varieties of Wisconsin cheese, sausages, cheese curds, ice cream, full menu, soda, souvenirs, and gifts. Mail Order. Online Orders.

Cornellier Superstore

2970 Milwaukee Rd., Beloit
608.364.1900
cheese-r-us.com
Family-owned specialty shop offering Wisconsin-made products. Mail Order. Online Orders.

Crystal Creek Dairy House

W7790 Hwy. 33, Beaver Dam
920.887.2806
Family-owned restaurant and cheese store that makes its own ice cream. Large variety of cheese and sausage. Mail Order.

Decatur Dairy, Inc.

W1668 Hwy. F, Brodhead
608.897.8661
Award-winning Havarti, Dill Havarti, Muenster, Stettler Swiss, and Colby Swiss as well as many varieties of squeaky-fresh cheese curds. Retail Store, Mail Order. Tours. Online Orders.

Edelweiss Cheese Authentic Wisconsin

202 W. Verona Ave., Verona
608.845.9005
edelweisscheeseshop.com
Over 120 Wisconsin cheeses. plus local beer, wine, crackers, and accoutrements.

Edelweiss Creamery

W6117 Cty. Hwy. C, Monticello
608.938.4094
edelweisscreamery.com
Led by a Wisconsin Master Cheesemaker, only artisan cheese factory in the U.S. to manufacture traditional 180-lb. Emmentaler Swiss Cheese. Also offers Butterkäse, Lacy Swiss, Havarti with Dill, and Muenster. Tours by appointment.

Ehlenbach's Cheese Chalet, Inc.

4879 Cty. Rd. V, DeForest
800.949.4791
ehlenbachscheese.com
Over 160 different kinds of Wisconsin cheese plus many other Wisconsin-made meats, gourmet items, candies, wines, and beer. Mail Order. Online Orders.

Fromagination

12 S. Carroll St., Madison
608.255.2430
fromagination.com
Cut-to-order cheese shop specializing in small-batch cheesemakers throughout Wisconsin, plus honey, jams, crackers, wine, and more. Online Orders.

Gibbsville Cheese Co., Inc.

W2663 Cty. Hwy OO, Sheboygan Falls
920.564.3242
gibbsvillecheese.com
Fine Cheddar, Colby, Monterey Jack and two-tone (Monterey and Colby) cheese available. Retail Store. Observation Window. Mail Order. Online Orders.

33 Henning's Cheese, Inc.
20201 Point Creek Rd., Kiel
920.894.3032
henningscheese.com
Famous for mammoth Cheddars (weighing up to 12,000 pounds), warm, fresh curds, Mozza Whips, and string cheese. Also sells ice cream, wines, souvenirs, and more. Retail Store. Observation Window. Museum. Mail Order. Online Orders.

34 Jim's Cheese
410 Portland Rd., Waterloo
800.345.3571
jimscheese.com
Large selection of award-winning Wisconsin specialty, artisan, and farmstead cheeses, including cheese aged up to 11 years. Also offering 250 cheese cutout shapes. Retail Store. Mail Order. Online Orders.

35 Knaus Cheese
N5722 Cty. Rd. C, Rosendale
920.922.5200/800.236.5200
stardairy.com
Wide variety of Weyauwega Star Dairy cheeses and holiday cheese boxes. Retail Store. Mail Order. Online Orders.

36 Kraemer Wisconsin Cheese
1173 N. 4th St., Watertown
920.261.6363
kraemercheese.com
Over 90 varieties of cheese plus Apricot Honey Cold Pack Cheese Spread and the 2010 World Champion Cheddar with Beer Cold Pack. Retail Store. Mail Order. Online Orders.

37 Krohn Dairy Store/Agropur Inc.
N2915 Cty. Rd. AB, Luxemburg
920.845.2901
World Champion Mozzarella and Provolone cheeses. Fresh curds on Mondays and Fridays. Retail Store. Mail Order.

38 Laack Cheese Co., Inc.
7050 Morrison Rd., Greenleaf
920.864.2815
Unique cheese spreads and large variety of specialty-flavored Cheddars and Jacks plus Aged Cheddars, string, whips, curds, and more. Retail Store. Mail Order.

39 Lamers Dairy
N410 Speel School Rd., Appleton
920.830.0980
lamersdairyinc.com
Wide variety of Wisconsin cheese and Wisconsin-made products including sausage, wine, maple syrup, honey, and much more. Retail Store. Observation Window. Mail Order. Online Orders.

40 Larry's Market
8737 N. Deerwood Dr., Brown Deer
414.355.9650
larrysmarket.com
Specialty food store offering Wisconsin artisan cheeses and gourmet foods. Retail Store.

41 Maple Leaf Cheese and Chocolate Haus
554 1st St., New Glarus
608.527.2000/888.624.1234
mapleafcheese and chocolatehaus.com
Quaint Swiss Chalet-style store featuring 100 varieities of Green county cheese, fudge, ice cream, and more. Mail Order.

42 Maple Leaf Cheese Store
W2616 Hwy. 11/81, Juda
608.934.1237
mapleleafcheesestore.com
Over 140 different varieties from Maple Leaf Co-op and other Green County cheesemakers. Retail Store. Mail Order.

43 Market Square Cheese
1150 Wisconsin Dells Pkwy. S., Lake Delton
608.254.8388
marketsquarecheese.com
Over 100 varieties of Wisconsin cheese plus candies, ice cream, and souvenirs. Mail Order. Online Orders. Tours.

44 Mars' Cheese Castle
2800 W. Frontage Rd., Kenosha
262.859.2244/800-655-6147
marscheese.com
Historic landmark castle offering over 300 cheeses, Cheddar bread, Wisconsin wines and beers, and sandwich counter. Mail Order. Online Orders.

45 **Mousehouse Cheesehaus**
4494 Lake Circle, Windsor
800.526.6873
mousehousecheese.com
Wisconsin's finest cheese and sausages,
including house-aged Cheddar and
Mousehouse Jack, plus wine, beer, deli,
and gifts. Mail Order. Online Orders.

46 **Noble View Creamery, LLC**
20715 Durand Ave., Union Grove
262.878.0619
nobleviewcreamery.com
Artisan cheeses made from farmstead
milk, including Juustoleipa and
Hispanic varieties. Retail Store. Mail
Order. Online Orders.

47 **Old Fashioned Foods Cheese**
331 S. Main St., Mayville
920.387.7924
oldfash.com
Cold pack spreads and process items,
plus over 40 hard cheeses from local
factories. Retail Store. Mail Order.

48 **Old Tavern Food Products, Inc.**
230 S. Prairie Ave., Waukesha
262.542.5301
oldtaverncheese.com
Wisconsin cheese and sausage
including 11 flavors of cheese spreads.
Custom corporate gifts available. Mail
Order. Online Orders.

49 **Otter's Lake Wisconsin Country Store**
N3280 Hwy. J, Poynette
608.721.3808
ottersstore.com
Wisconsin cheese and foods, fine beers,
hand-dipped ice cream, and gifts in a
log cabin setting. Mail Order. Online
Orders.

50 **Paoli Cheese**
6890 Paoli Rd., Belleville
608.845.7031/800.762.4644
paolicheese.com
Cheese, sausages, jams, and snacks.
Mail Order. Online Orders.

51 **Pine River Dairy**
10115 English Lake Rd., Manitowoc
920.758.2233
pineriverdairy.com
A butter manufacturing plant with
a cheese retail store offering 250
varieties. Observation Window. Mail
Order. Online Orders.

52 **Plymouth Wine and Cheese**
3240 Cty. Hwy. PP, Plymouth
920.892.8781
plymouthwineandcheese.com
Cheese, sausage, dips, wine, beer, and
more. Online Orders.

53 **Roelli Cheese Haus**
15982 Hwy. 11, Shullsburg
608.965.3779/800.575.4372
roellicheese.com
Small batch, hand-made Cheddar curds
and specialty Cheddar Blue cheeses.
Retail Store. Observation Window.
Mail Order. Online Orders. Tours by
appointment.

54 **Ron's Wisconsin Cheese, Inc.**
124 Main St., Luxemburg
920.845.5330
ronscheese.com
Cheddars from mild to 10-year. Retail
Store. Mail Order. Online Orders.

55 **Roses Fresh Market**
433 Broad St., Lake Geneva
262.248.8168
Local cheese, dairy, produce, butcher-
quality, grass-fed meats. Fresh wild fish
flown in weekly, and more.

56 **Sassy Cow Creamery**
W4192 Bristol Rd., Columbus
608.837.7766
sassycowcreamery.com
State-of-the-art farmstead creamery
and retail store with Sassy Cow milk,
heavy cream, ice cream, etc. as well as
large variety of Wisconsin cheese and
other products. Observation Window.

57 **Saxon Creamery**
855 Hickory St., Cleveland
920.693.8500
saxoncreamery.com
Homestead cheese made from the milk
of one herd that grazes near the shores
of Lake Michigan. Retail Store.

58 Say Cheese
117 S. Royal Ave., Belgium
262.285.3036
saycheesewi.net
Over 150 varieties of Wisconsin cheese, plus many other Wisconsin products, deli, and a grilled cheese bar. Retail Store. Online Orders.

59 Schultz's Cheese Haus
N6312 Hwy. 151 S., Beaver Dam
920.885.3734/800.236.3734
schultzscheese.com
Over 135 varieties of Wisconsin cheese, plus sausage, fine wines, gift boxes, and more. Mail Order. Online Orders.

60 Silver-Lewis Cheese Factory Co-op
W3075 Cty. Rd. EE, Monticello
608.938.4813
Small, historic factory producing award-winning Brick, Muenster, Farmer, and flavored Farmer varieties. Retail Store. Mail Order.

61 Simon's Specialty Cheese/Agropur
2735 Freedom Rd., Appleton
920.788.6311
simonscheese.com
Award-winning Cheddar, mozzarella, Provolone, feta, and more, plus Wisconsin wines/beer and local specialty foods. Retail Store. Mail Order. Online Orders.

62 Specialty Cheese Co., Inc.
430 N. Main St., Reeseville
920.927.3888
specialcheese.com
Specializing in Mid-Eastern, Indian, Hispanic, low-carb, and gluten-free cheese. Retail Store. Online Orders.

63 Tenuta's Delicatessen
tenutasdeli.com
3203 52nd St., Kenosha
262.657.9001
Italian market and Kenosha tradition since 1950, featuring specialty groceries, meats, cheeses, and prepared food. Online Orders.

64 The Brat Stop
12304 75th St., Kenosha
262.857.2011
bratstop.com
Brats, restaurant, cheese mart, & banquet facilities. Mail Order. Online Orders.

65 The Cheese Nook
8133 Hwy. 10, Branch/Whitelaw
920.684.3302/800.736.5220
Cheese, gift boxes, baskets, and trays.

66 The Cheese People of Beloit, LLC
431 E. Grand Ave., Beloit
608.207.3094
thecheesepeople.com
Cut-to-order artisan cheese, craft beers, and wines. Mail Order. Online Orders.

67 The Elegant Farmer
1545 Main St., Mukwonago
262.363.6770
elegantfarmer.com
Farm kitchen bakery, deli, and cheese shop complex. Online Orders.

68 Tim and Tom's Cheese Shop and More
12009 53rd Pl., Kenosha
262.857.4606
timandtomscheeses.com
Hundreds of varieties of Wisconsin cheese, plus honey, mustards, jams, maple syrup, and more. Mail Order. Online Orders.

69 Union Star Cheese
7742 Cty. Rd. II, Fremont
920.836.2804
unionstarcheese.com
Cheddar, Colby, string cheese, fresh curds, and more. Retail Store. Mail Order. Online Orders. Tours.

70 Vern's Cheese, Inc.
312 W. Main St., Chilton
920.849.7717
vernscheese.com
Wisconsin cheeses, meats, wines, glass-bottled milk, ice cream, and more. Mail Order. Online Orders.

West Allis Cheese and Sausage Shop
wacheese-gifts.com
Over 200 varieties of cheese plus
Wisconsin sausages, brats, jams, jellies,
honey, and spreads. Three locations:

71A 6832 W. Becher St., **West Allis**
414.543.4230

71B 400 N. Water St., **Milwaukee** (at
the Milwaukee Public Market)
414.289.8333

71C Oakland Cafe and Deli
2974 N Oakland Ave., **Milwaukee**
414.962.5455

72 Weyauwega Star Dairy
109 N. Mill St., Weyauwega
920.867.2870
wegastardairy.com
Specializing in Italian cheeses. Retail
Store. Mail Order. Online Orders.

73 Widmer's Cheese Cellars
214 W. Henni St., Theresa
888.878.1107
widmerscheese.com
Handcrafted, award-winning Brick
made the old-fashioned way, Cheddar,
and real Wisconsin Colby, as well as 60
other varieties of cheese and sausage.
Retail Store. Observation Window. Mail
Order. Tours.

74 Willow Creek Cheese
W1965 St. Rd. 21, Berlin
920.361.5265
unionstarcheese.com
Curds, string cheese, blue, Brick,
Cheddar, and more. Retail Store.
Observation Window, Mail Order. Tours.

75 Wisconsin Cheese Mart
1048 N. Old World Third St.,
Milwaukee
888.482.7700
wisconsincheesemart.com
Large selection of Wisconsin cheese.

76 Wisconsin Cheese and Wine Chalet
11190 Goede Rd., Edgerton
608.884.3270/888.6CHALET
wisconsinchalet.com
Free cheese and wine tasting daily, full
deli, outdoor patio. Over 100 Wisconsin
cheeses, plus microbrews, sausages,
salsas, mustards, and other specialties.
Online Orders.

77 Z's Wisconsin Cheese Store
807 Jay St., Manitowoc
920.652.9980
zswisconsincheesestore.com
Cheese and wine from around
the world. Fresh cheese curds on
Wednesdays and Fridays.

78 Zimmerman Cheese, Inc.
N6853 Hwy. 78, South Wayne
608.968.3414
Award-winning Baby Swiss, Muenster,
and fresh Muenster Curds, plus a wide
variety of cheeses for sale. Retail Store.
Mail Order.

0 33441 99264 1

$1.00 off

Good at the Dane County Farmers' Market

Hook's Cheese Company, Inc.

World Champion Cheese Makers

Tony & Julie Hook • Mineral Point, Wis.

www.hookscheese.com

Exp. 12/31/2014

Receive 10% off

Marieke® Gouda
&
Marieke® Golden Cheeses Only

Redeem in our retail store in Thorp, WI

Expires 12/31/2014

N13851 Gorman Ave., Thorp, WI 54771

Marieke® Gouda

Penterman Farm
HOLLAND'S FAMILY CHEESE LLC

CREAMERY/STORE
715-669-5230

HollandsFamilyCheese.com

10% OFF YOUR TOTAL PURCHASE*

ON THE CAPITOL SQUARE

fromag|nation

artisanal cheeses & perfect companions TWELVE SOUTH CARROLL, MADISON, WI 53703 • FROMAGINATION.COM

Wine & Cheese Tasting

Four Wisconsin Cheeses
& Four Wine Pairings
$30 all inclusive

GRAZE

from local pastures

on the Capitol Square
1 S Pinckney Street, Madison, WI
608.251.2700 ~ grazemadison.com

Valid Monday-Thursday during regular service hours.
Limit one coupon per table per transaction. Expires 12/31/2014

$1.00 off

**any RP's fresh
or frozen pasta
product**

**RP's
PASTA
COMPANY
MADISON WI**

Valid only at the Dane County Farmers Market Expires November 8, 2014

Try some of our many award winning cheeses, or one of our new blue cheeses:

Hook's Barneveld Blue - a tangy goat milk blue.

Hook's "1472" Blue - "EWE CALF to be KIDding" - a mixed milk blue cheese with sheep, cow, and goat milk."

Expert, licensed cheesemaker, Marieke Penterman and her team, handcraft traditional Dutch Goudas using the time-tested, Old World, cheesemaking methods Marieke brought with her when she emigrated from the Netherlands.

Marieke transforms farmstead-fresh, raw, cow's milk from her family farm into award-winning cheese in the heart of Central Wisconsin. Each batch is then carefully cured on imported Dutch pine planks in temperature and humidity-controlled aging cellars.

Marieke was awarded Best in Show at the 2013 U.S. Championship Cheese Contest for Marieke® Mature Gouda, aged 6-9 months.

From freshly made bakery to late night Happy Hour specials, GRAZE serves creative, casual dishes, made from seasonal and local ingredients year round.

Join us for our Local Beer Dinners, seasonal patio seating, and Friday night fish fry -- all with a spectacular view of the Capitol.

1 S Pinckney Street, Madison, WI ~ 608.251.2700 ~ grazemadison.com

"Authentically delizioso." —OPRAH MAGAZINE

RP's Pasta Company's products are made with the finest all-natural ingredients, an appreciation for Old World culinary traditions, and a love for food.

The Wisconsin Milk Marketing Board

Excerpted and adapted from eatwisconsincheese.com

The Wisconsin Milk Marketing Board (WMMB), headquartered in Madison, is a non-profit organization established in 1983 by Wisconsin dairy farmers and funded entirely by Wisconsin's dairy farm families. For every 100 pounds of milk produced and marketed in the state, ten cents goes to the WMMB to promote Wisconsin-produced dairy products. An additional five cents goes to the National Dairy Promotion & Research Board for generic dairy promotion activities at the national level.

Farmers also run the WMMB. Twenty-five dairy farmers, elected by their peers, serve on the Board for three-year terms. The Board's farmer-directors have a direct involvement in planning and monitoring the organization's marketing and promotional programs, which are conducted by a staff of marketing, research, and communications professionals.

The mission of the WMMB is to increase the sale and consumption of Wisconsin milk and dairy products by providing programs that enhance the competitiveness of the dairy industry. The WMMB promotes Wisconsin dairy products at the state and national levels through a comprehensive array of promotional resources, programs, and merchandising tools, including the following:

▸ Regional marketing managers work with Wisconsin companies and national grocery and foodservice operators and distributors to ensure that the more than 600 varieties, types, and styles of Wisconsin cheese and other dairy products from America's Dairyland are available and promoted in all major U.S. markets—to more than 300 million consumers.

▸ The WMMB supports the Wisconsin Center for Dairy Research at UW-Madison, a world class dairy research and applications facility that works closely with Wisconsin manufacturers to make sure that the cheese and dairy products meet the highest standards for quality and safety.

▸ More than 60 County Dairy Leader Groups receive WMMB support to help them showcase and promote our dairy industry in local communities across Wisconsin.

▸ The nutritional education arm of the WMMB, the Wisconsin Dairy Council, has six regional program managers throughout Wisconsin who work with school foodservice staff and teachers to promote the nutritional benefits of dairy products for students.

8418 Excelsior Dr.
Madison, WI 53717
(608) 836-8820

info@EatWisconsinCheese.com
eatwisconsincheese.com

Organic Valley and Kickapoo Blue Cheese

Based in La Farge, Wisconsin and founded in 1988, Organic Valley is the largest cooperative of organic farmers in the United States. Here, Organic Valley introduces one of the newest products in its line-up of more than 30 kinds of cheese.

Organic Valley Kickapoo Blue Cheese is a local inspiration. Our family farm cooperative was born and raised in the lush Kickapoo Valley of southwest Wisconsin. This is home. It is where we began and continue every day our all-good mission to save family farms by making delicious food in harmony with nature.

One-part pasture, one-part woodland, and one-part paradise, this valley is also home-sweet-home to sassy cattle, darting trout, and bumbling bees. Truly a land of milk and honey—and cheese. How could we not try our hand at creating a soul-soaring cheese? Maybe it wasn't as much an inspiration as it was a no-brainer. We just knew this cheese would be a special treat.

With three first-place prizes at the World Dairy Expo and too many blue ribbons to count, there's no denying the tasty result. Kickapoo Blue is a richly marbled, crumbly cheese (aged 60 days minimum) with an earthy, delectable flavor. No wonder Kickapoo Blue has such a long line of dance partners, including pears, apples, walnuts, red wines, and Port. It's delicious in salads, dips, sauces, and spreads, or with veggies and fruits.

Like all Organic Valley cheese, the secret ingredient is the ample amount of time our cows spend munching organic pastures. Our family farmers are pioneers of ages-old grazing practices that replenish soil and create the world's best milk. All the nutritional goodies—calcium, conjugated linoleic acid (CLA), omega-3, carotene—and dairy-fresh taste are guaranteed when cows eat what they are meant to eat, out in the fresh air and nourishing sunshine.

Of course the taste of Kickapoo Blue is divine: It's from organic pastures at the heart of Wisconsin.

Try Kickapoo Blue in the following recipes: Polenta with Blue Cheese and Caramelized Onions (page 56); Papparadelle with Zucchini, Blue Cheese, and Tarragon (page 92); and Bacon, Blue Cheese, and Pear Bites (page 118). And check out the Cheese of the Month feature about blue-veined cheese on page 66.

www.organicvalley.coop

Photos courtesy of Organic Valley.

About Wisconsin Master Cheesemakers

©Wisconsin Milk Marketing Board, Inc.

Crafting cheese is taken very seriously in Wisconsin. In 1994 Wisconsin began its three-year Master Cheesemaker Program, a rigorous, advanced education program for experienced cheesemakers. It is jointly sponsored by the Wisconsin Center for Dairy Research, UW-Extension, and the Wisconsin Milk Marketing Board. Outside of Europe, there are no other comparable programs.

Requirements for admittance into the Wisconsin Master Cheesemaker Program are stringent. An applicant must be an active licensed cheesemaker for a minimum of 10 consecutive years and work in a Wisconsin-licensed cheese plant before being accepted into the program. Applicants must have been making the specific variety of cheese for which they desire certification for at least five years.

Candidates for Master Cheesemaker status must complete a three-year course of study that includes a formal sequence of four required and three elective courses. Participants also enroll in an apprenticeship program for the specific cheese(s) for which they are seeking certification. They may seek certification for no more than two varieties of cheese per each three-year program. During each year of the three-year apprenticeship, cheesemakers must submit samples of these specific cheeses for evaluation by a designated committee for quality and consistency. After the course work and apprenticeship requirements are met, cheesemakers take a written exam.

Master Cheesemaker status is very beneficial to cheesemakers who complete the program. Graduates obtain a thorough understanding of the science of cheese making and perfect their already considerable cheese making skills. Master Cheesemakers become highly competitive in Wisconsin, the U.S., and abroad, and the cheese they craft takes on added value. Look for the special label—the blue Master's Mark Wisconsin label—that is placed on packages of cheese for which a cheesemaker has received Master Cheesemaker certification.

With the addition of the 2013 graduates, a total of 56 Wisconsin cheesemakers representing 31 companies hold Master Cheesemaker certification for a total of 33 varieties of cheese. Some cheesemakers have participated in the program more than once and have received certification in additional cheese varieties. The most certificates obtained by a single cheesemaker is nine (Bruce Workman, Edelweiss Creamery, Verona).

REAP FOOD GROUP

Nourishing the links between land and table.

www.reapfoodgroup.org

REAP makes connections that help grow a better food system for southern Wisconsin.

About REAP: Moving Good Food Forward

A not-for-profit organization based in Madison, REAP Food Group nourishes the links between land and table to grow a healthful, just, and sustainable local food system. REAP staff, members, and volunteers envision a time when there is sustainable, local food on every plate.

REAP is committed to projects that:

> ‣ shorten the distance from farm to table

> ‣ support small family farmers

> ‣ encourage sustainable agricultural practices

> ‣ preserve the diversity and safety of our food supply

> ‣ address the food security of everyone in our community

Farm to School

REAP's Farm to School program teaches the next generation of eaters to love healthy food from local farms by providing access to fresh, local, and sustainably-produced foods for Wisconsin school children in their lunchrooms and classroom. During the 2012-2013 school year, REAP purchased more than 9 tons of produce from local, sustainable farms, feeding 4,500 students a week. REAP Farm to School has served more than a half a million healthy local snacks since 2006.

Buy Fresh Buy Local

REAP's Buy Fresh Buy Local program connects farmers, chefs, and consumers to build lasting relationships that provide economic benefits to all parties. The Buy Fresh Buy Local program is made up of 42 restaurants, four health care organizations, and two grocery stores that continually show their commitment to local, sustainable food and are leading the way in building our community of local eaters. Since 2006, partner restaurants have purchased $13 million in local goods directly from local producers.

Farm Fresh Atlas

The *Southern Wisconsin Farm Fresh Atlas* is a go-to source for finding farms, businesses, restaurants, and farmers' markets that sell their goods directly to customers in our community. The *Farm Fresh Atlas* has been building the market for local, sustainable food since 2002.

REAP Events in 2014

Spring Gala
Sunday, April 27, 2014
Madison Concourse Hotel Grand Ballroom, 1 West Dayton St., Madison

At this elegant dinner event, chefs prepare a unique, formal five-course dinner at tableside. Local foods are the focus of each course, with wine pairings to complement each chef's menu.

Burgers & Brew
Saturday, June 7th, 2014
Capital Brewery Bier Garten, 7734 Terrace Ave., Middleton

Join the party when some of Dane County's most popular chefs pair up with Wisconsin's talented brewers and farmers to serve up unique, mouthwatering burgers and satisfying craft beers.

Food for Thought Festival
Saturday, September 20th, 2014
Martin Luther King Jr. Blvd. (off the Capitol Square), Madison

The annual Food for Thought Festival is a fun, festive forum that explores and celebrates our many opportunities to eat more pleasurably, healthfully, and sustainably.

Pie Palooza
Sunday, November 2, 2014
Goodman Community Center, 149 Waubesa St., Madison

Enjoy a scrumptious brunch showcasing sweet and savory pies by Madison-area chefs using seasonal, local ingredients. Want to eat locally? It's "as easy as pie"!

Local Harvest Dinners
Multiple Dates
Buy Fresh Buy Local partner restaurants, Madison area

Local Harvest is a celebration of local foods and the chefs who turn them into delicious meals. Each participating restaurant features special menu options or events.

About the Authors

Photo by Susan Chwae

Joan Peterson is the publisher and creator of the award-winning EAT SMART series, which includes guides to the cuisine of Brazil, Turkey, Indonesia, Mexico, Poland, Morocco, India, Peru, Sicily, France, Norway, and Germany. Each book has been designed for food lovers who want to navigate menu and market with confidence. She co-authored with Terese Allen the previous editions of the *Wisconsin Local Foods Journal*. Joan leads culinary tours to Europe, Asia, and Africa, and is a founding member of the Culinary History Enthusiasts of Wisconsin (CHEW). Her publishing company, Ginkgo Press, is in Madison, Wisconsin. Visit her website at eatsmartguides.com.

Photo by Jim Block

Terese Allen has written scores of books and articles about Wisconsin's culinary culture, past and present. A columnist for *Edible Madison* and *Edible Door* magazines, she is also president of the Culinary History Enthusiasts of Wisconsin (CHEW) and a longtime director of REAP Food Group. Her award-winning titles include *The Flavor of Wisconsin* and *The Flavor of Wisconsin for Kids*, and she's currently working on a memoir about eating local. Terese lives in Madison and on Washington Island, and is hungry all the time. Visit her website at tereseallen.com.

Wisconsin Local Foods Journal Order Form

Please send me _______ copies of the 2014 *Wisconsin Local Foods Journal.*

Name____

Address_________________________________ Phone (_____)__________

City_________________________________ State_____Zip_____________

Email____

___ copies of the 2014 Wisconsin Local Foods Journal x $17.95 $ ______

     U.S. Shipping: $6.00 + $1.50 each additional calendar $ ______

     Wisconsin residents add 5% sales tax $ ______

     Total amount enclosed $ ______

Send check/money order payable to Ginkgo Press to: PO Box 5346, Madison, WI 53726. OR order online at www.wisconsinlocalfoodsjournal.com.

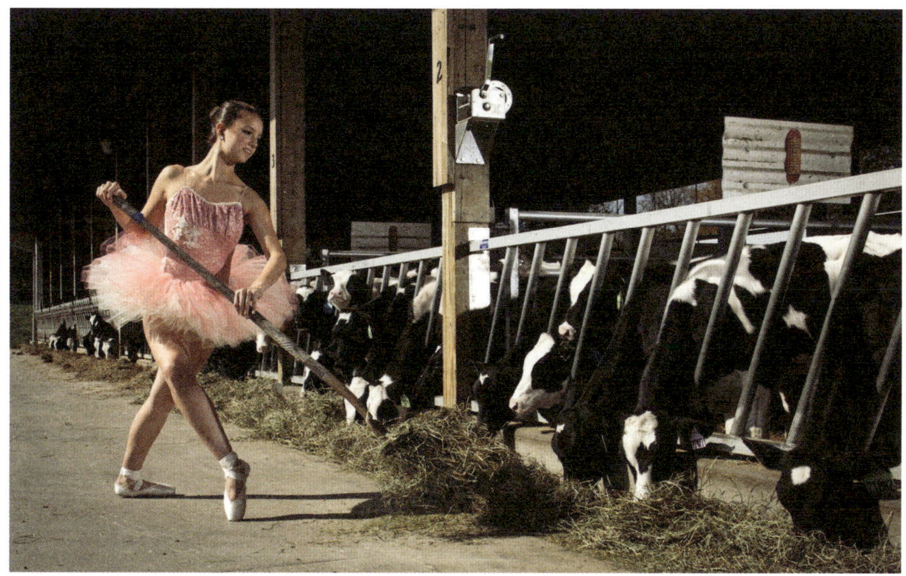

Wisconsin's cows appreciate the arts. ©Becky McKenzie Photography.

Index

Recipes are in *italics*; photos are in **bold letters**.